The YOGA *of* TIME TRAVEL

Other Books by Fred Alan Wolf

Matter into Feeling

Mind into Matter

The Spiritual Universe

The Dreaming Universe

The Eagle's Quest

Parallel Universes

The Body Quantum

*Star*Wave*

Taking the Quantum Leap

Space-Time and Beyond: The New Edition
(with Bob Toben)

Praise for Fred Alan Wolf

The Yoga of Time Travel *is one of the most imaginative books I have read about the nature of time. It makes us wonder if time travel is possible, not only through the use of technology, but through yoga, which anyone can practice.*

—AMIT GOSWAMI, PH.D., author of *Physics of the Soul*

In **The Yoga of Time Travel**, *Fred Alan Wolf takes the reader beyond the normal boundaries of space and time into the world of the infinite, eternal spirit. He blends physics and spirituality in a way that is both seamless and intriguing.*

—GLEN KEZWER, author of *Meditation, Oneness, and Physics*

Wolf is at the forefront of creating a new integral science that includes psychology, physics, and spiritual thinking. In **The Yoga of Time Travel**, *he makes concepts such as black holes, the space-time continuum, and parallel universes something we can feel.*

—ARNOLD MINDELL, PH.D., author of *The Quantum Mind and Healing*

Fred Alan Wolf's **Mind into Matter** *breaks new ground that is both enthralling and energizing, giving us maps to explore consciousness and the cosmic mystery. This is quantum alchemy at its finest.*

—MICHAEL TOMS, cofounder and host of *New Dimensions Radio*

Entering Wolf's **The Spiritual Universe** *is like stepping into a laboratory in a spaceship careening through the universe and manned by a philosopher-scientist. If you have a head for scientific reasoning, you'll enjoy the spirited journey. If you're a romantic, earthbound, poet-type like me, you're in for some mind expansion.*

—THOMAS MOORE, author of *The Reenchantment of Everyday Life*

The Spiritual Universe *creates a brilliant New Physics of the Soul and ushers us into the third millennium with deepening faith, bold heart, and profound insight.*

—ERNEST L. ROSSI, PH.D., author of *The Psychobiology of Gene Expression*

Fred Alan Wolf's **Parallel Universes** *is a wild intellectual ride—an enthralling read.*

—PUBLISHERS WEEKLY

The YOGA of TIME TRAVEL

How the Mind Can Defeat Time

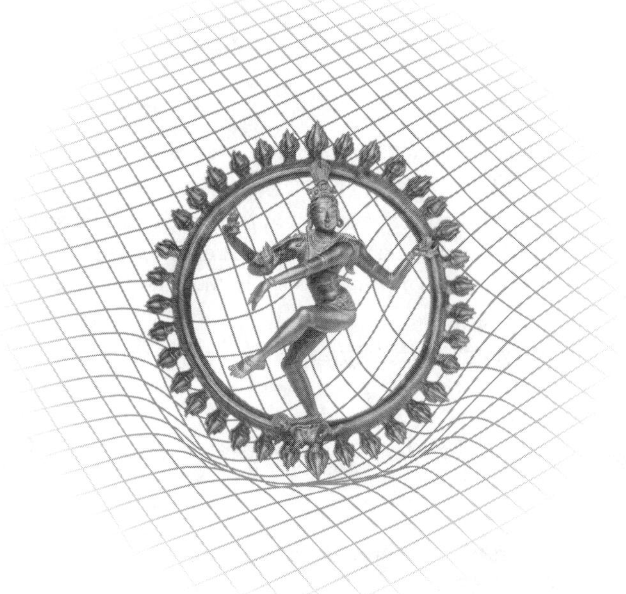

FRED ALAN WOLF, PH.D.

Wheaton, Illinois ♦ Chennai (Madras), India

First Quest Edition 2004

Quest Books
The Theosophical Publishing House
P. O. Box 270
Wheaton, IL 60189-0270
www.questbooks.net

Cover art, book design, and typesetting by Dan Doolin

Picture Credits:

Page 17: Illustration used with permission courtesy of The Bhaktivedanta Book Trust Int'l © 2004.

Page 58: Photo used with permission courtesy of Jerry Davidson, Webmaster@CosmicHarmony.com. From website http://www.cosmicharmony.com/Sp/Ramana/Ramana.htm.

Library of Congress Cataloging-in-Publication Data

Wolf, Fred Alan.
The yoga of time travel: how the mind can defeat time / Fred Alan Wolf.—1st Quest ed.
p. cm.
Includes bibliographical references and index.
ISBN 0-8356-0828-X
1. Space and time. 2. Yoga. 3. Time travel. 4. Time—Psychological aspects. 5. Time—Religious aspects. I. Title.

QC173.59.S65W65 2004
115—dc22

2004044782

5 4 3 2 1 * 04 05 06 07 08 09

Printed in the United States of America

To my future readers:
May your travels be timely

CONTENTS

ACKNOWLEDGMENTS

I would like to thank Carolyn Bond for her skillful editing of the manuscript and for making several important improvements in the writing. To my editor in chief at Quest Books, Sharron Dorr, I offer my gratitude for her editing and for offering many valuable suggestions. I also want to thank my wife Sonia Sierra Wolf, who—being a student, a teacher, and practitioner of Iyengar yoga—made numerous helpful contributions.

INTRODUCTION

"I don't understand you," said Alice. "It's dreadfully confusing!"
"That's the effect of living backwards," the Queen said kindly: "it always makes one a little giddy at first—"
"Living backwards!" Alice repeated in great astonishment. "I never heard of such a thing!"
"—but there's one great advantage in it, that one's memory works both ways."
"I'm sure mine only works one way," Alice remarked. "I can't remember things before they happen."
"It's a poor sort of memory that only works backwards," the Queen remarked.

—Lewis Carroll,
Through the Looking-Glass

Most of us assume, as Alice does, that whatever we can remember has already taken place. If asked why we don't remember scenes from our future, we might answer: "Because, dummy, they haven't happened yet!" But as the Queen in Lewis Carroll's delightful book suggests, perhaps we *do* have memories of the future, however nonsensical that may sound. Consider the albeit radical possibility that the Queen is right: memory *does* work both ways. That is, you are perfectly able to remember the future just as well as you can recall the past. Further, consider that having a two-way memory could lead, as the Queen suggests, to distinct advantages. For example, it might help you deal with synchronicities and experiences of *déjà vu,* avoid health problems, make significant predictions about your life, and offer many other benefits, as may become clear as this book unfolds.

To begin exploring this idea, let's think first about the nature of memory as we commonly know it—having to do with the recall of past events. Certainly you remember your last vacation, as well as a favorite restaurant you went to, or a show you saw, and so on. And I'm sure there are some past experiences you don't remember, though possibly your spouse does: "Oh, don't you remember that day in Paris when we saw those flowers on the bank of the Seine?" she or he asks, and you draw a blank.

Ever wonder why your companion remembers things that you don't? The popular conception, based on brain research,[1] is that whether you recall any details or not, your memory contains a complete record of your past, as if it were a movie. You are most likely, however, to recall only those events that made an impression on you. That day in Paris, the problem was that you simply weren't paying attention, and those flowers along the Seine affected your spouse more deeply than they did you.

To be sure, sometimes we also forget events that *have* made a great impression. Usually they have been traumatic, and we don't *want* to remember them. In some such cases, though, deep psychoanalysis can help us improve our recall.

Regardless, further analysis by memory experts indicates that the popular adage is false: Memory is not restricted to only what has made an impression on us, positively or negatively. To the contrary, it turns out that actively working on one's memory can greatly enhance it. And it turns out this work can lead to remembering, not only the past, but also the future. As we will discover, this effort plays a key role in the mind-yoga that allows for time travel.

So, suppose you had been to the future and what you saw was either so uneventful that you didn't notice or so scary that you simply decided *not* to remember it. According to what we shall find out in this book, your ability to remember the future depends on your ability to pay close attention to these future events, not just idly glance over them as you may have done the flowers along the Seine. With some guidance and analysis, perhaps you could learn to recall the future with as much success as such procedures can enable you to recall forgotten past events.

I've heard that some therapists use a technique called "past-life recall" to help patients deal with unexplained trauma and psychological problems they are encountering in this life. I have also heard of a technique that enables people to "recall" future lives or events so they are better prepared to face what seems inevitable or unavoidable in the near or distant future. Whether this is pure imagination or wishful thinking is difficult to say. Of course if you only believe in the present moment—whatever that may be—such a discussion seems pointless and perhaps unscientific. But suppose there were a reasonable scientific basis for believing in the concrete existence of both the past and the future—coexisting with the present in some yet to be determined manner. Then what? In that case, both the future and the past would be as real to you as the drugstore on the corner or the North Pole, even though—sitting in your chair reading this book—you aren't at either of those places now. You certainly wouldn't remember the North Pole if you hadn't been there yet, would you? But that doesn't mean the North Pole does not exist. By the same token, perhaps the future is just as real, and the only reason we have no memory of it is because we haven't visited it yet.

But let's suppose you had "been there and done that," as they say. What would it mean to have a memory of the future? Isn't memory a record of what you did in the past? But if in the "past" you went to the future, how would you deal with a memory of it? Trying to think this way does make one, as the Queen puts it, "a little giddy at first."

Indeed, such ideas may seem like science fiction, but when we examine what scientists are doing these days in terms of realizing time travel and time manipulation, you will see that science fiction has become science fact. My hope is that if nothing else, after reading this book you will understand just what is meant by time travel and why scientists are now taking it seriously.

Surprising as it may seem, a scientific basis for time travel was established more than a hundred years ago; Herbert George Wells wrote about it in 1895, and Albert Einstein and Hermann Minkowski showed how it was theoretically possible in 1905 and 1908. In fact, more than fifty years ago, scientists were proving time travel to be a reality. Documentation shows that in carefully defined laboratory experiments, objects were observed that literally slowed down in time, such that some of them lived nine or ten times their natural life span.[2]

Sounds unbelievable? I'll explain more about that experiment shortly. In the meantime, let me tell you a secret: Some of the remarkable people you meet in life are time travelers. A few of these people know it; the others time travel without realizing it, but they do it just the same. These are the people who appear older than their years or, yes, often enough considerably younger. I, too, time travel. In fact, I do it nearly every day, especially when I find myself in creative activity—lost in my work, as we say. Later, we'll look more deeply into this phenomenon, too.

SOME PRELIMINARY MEANDERINGS IN THE TEMPORAL STREAM

Just think what it would mean to live nine or ten times longer than your putative four-score-plus years—that is, perhaps as long as eight hundred years! Or imagine that you live through ten years of time while those around you only experience one second of time passing, or that you experience one second of time passing while those around you age ten years.

In the latter case, during those ten years each of them would experience the earth daily rotating about its axis and note its yearly movement across the solar system, but you would not. Traveling through time at this breakneck "speed," you would grow one day older while the world around you ages more than 86 thousand years. In ten years of your life lived at this rate, nearly countless generations of humanity would age more than 315 million years—enough time for you to see evolution on a scale beyond imagination.

The former case would be equally strange, since the world and all of its processes would slow down terribly, so much so that the world around you would grow strangely silent, dark, and still. Even light would move very slowly from your point of view. Light travels at more than 670 million miles per hour, but that hour would stretch out for you to 36 thousand years, slowing light down to a crawl of about two miles per hour for you. You can walk faster than that! Since you wouldn't see light until it struck your eyes, you would experience the world in flashes, like a stroboscopic light show.

However, even this scenario isn't the whole story. It assumes that you could hold on to the normal timing of your own bodily processes and think as you normally do, with full neuronal cooperation at your normal speed of functioning. But if your body's processes slow down as well, things would get even more interesting. Consider your sense of sight. If the speed of light slowed down, so would its vibrational rate, which means that colors would change so drastically that they would be impossible to see with your eyes. A similar slowing of all of the physical phenomena around you would result. In other words, the world would most likely vanish from your senses if you were aging ten years in one second.

Even more bizarreness awaits the time traveler who can move *backward* through time. New paradoxes pop up, depending on who moves relative to whom. If, for example, you move backward through time while the world around you passes at the normal rate of one second per second into the future, you will gradually get younger while those around you age. If you move backward into time even faster, you run into the paradox of just what happens to you when you reach the moment of your birth. Do you then need your mother to be present? Even worse, suppose you move to the period just before the sperm meets the egg that made you. Since you wouldn't be a "you" yet, just what would be going on? What would happen to your consciousness in a time before your conception?

Or consider the other possibility: You move backward through everyone else's time stream so that while you see them grow younger and all processes running backward in time—like a movie in reverse—you go on aging at a normal rate. Perhaps in one second you move counter to a ten-year retrograde time stream. In one year you would move back more than 86 thousand years.

Is anything like this even possible? Suppose you went back more than 500 million years into the past, before humans even evolved. What would happen if you accidentally stepped on a life form that was one of your ancestors? Could you ever be born?

In this book we'll examine several such temporal paradoxes and I'll show you how it is possible, from the point of view of physics, to beat the paradox game and return to any point in time you wish without suffering any obvious consequences. I say "obvious" because even though there *are* consequences of time travel, they aren't what we might expect. As we shall see, it all has to do with the mind and learning to change possibility into reality.

Staying Young While Living Longer

Let's take a look at that experiment in which, more than fifty years ago, scientists observed objects that lived nine or ten times their expected life span.

Every day subatomic particles are created whenever cosmic particles from the sun or a distant galaxy collide with particles in our upper atmosphere. Specifically, these cosmic particles are protons, once known as cosmic rays, that are subatomic particles making up the nuclei of atoms. Few cosmic rays make their way to sea level. Hence nearly all of these newborn particles, called *muons* or *mu mesons,* are created at very high altitudes of our planet. These newborns can be counted with a little patience and a special device called a *scintillation counter* (which, as its name suggests, scintillates when something very tiny, like a muon, hits it). These counting devices can also determine what happens to these little babies after they have been detected. They can even count how long they live and what happens to them when they die. Upon death these particles decay, and when they decay, they suddenly disappear, leaving behind remnants.[3]

Whereas we humans have a life span of around eighty years, give or take a few, muons survive intact for a much briefer time—an average of about two microseconds (two millionths of a second). However, some die very quickly, in under one microsecond, and some live for as long as six microseconds. Very few are found at the end of, say, eight microseconds.

In one experiment, physicists took scintillation counters to the top of a mountain 6300 feet above sea level. They counted the number of muons at that altitude and found that somewhere around 568 newborns passed into their counters each hour. They then followed the muons through their short lives, letting them travel down a short vertical tube where they came to rest and eventually decayed near a second scintillation counter. As expected, only 300 resting muons lived past two microseconds. Around 30 of them made it to the ripe old age of 6.3 microseconds.[4] Because the scientists knew how far these particles traveled along the tube's length, they could determine how fast they flew before they rested and decayed, and they found that they moved at very near lightspeed.

Next, they took their counters down to the seaside. What did they anticipate there? Well, if a muon lived long enough and moved at near lightspeed, it could travel the 6300 feet down to sea level in about six microseconds. But given that most of them don't live that long, the scientists expected to find only a handful surviving—maybe 30 oldsters, say, who could make the journey.

Surprisingly, however, many more than 30 survived. In fact, around 412 made the trip without mishap.

How could that many live that long? Travel may add a certain pizzazz to one's life, but I have never heard of it lengthening one's life span. That is, not unless you take Einstein's relativity theory into account. The theory says that time does not function the same way for a moving object as it does for one standing still. Moving objects experience a slowing down of time, so that while the rest of the world passes through a given time period, the moving object passes a shorter time period. In this respect, we can estimate how long the 412 muons that reached sea level "thought" they had lived. It turns out that that they experienced a time period of only around 0.7 microseconds. Compare that with 6.3 microseconds—the time it takes to make the trip down the mountain at near lightspeed—and you see that this yields a factor of 9, exactly what would be calculated by Einstein's theory. In other words, the muons that survived the trip lived more than nine times their expected life span.

What is going on here? For the muons, nothing really extraordinary happened. They just lived their short, seven-tenths-of-a-microsecond life spans on their way down the mountain. But it just so happens that we on the ground passed through 6.3 microseconds of our life spans at the same time that the muons passed through only 0.7 microseconds. In what sense did these two periods take the same amount of time? In trying to think about such things, our very figures of speech become perplexing. Our language is so based on thinking in terms of *absolute* time that the mere idea of *relative* times hardly makes any sense. As Alice says, "It's dreadfully confusing!"

Relative *distances,* on the other hand, make sense. I can travel from my living room to my bedroom—some dozens of feet—by walking off a mile if I go downstairs, out the door, and around the block a few times before I walk into the bedroom. Or I can walk to the kitchen first and then to the bedroom. Each measure of distance is different. The distance is relative to the route I take. I always start in the living room and end up in the bedroom, but the distance I travel to get there can be, and is normally, different (since I rarely walk in a straight, shortest-distance line) each time I make the journey.

We assume that, in contrast to moving through space, moving from one point to another in time is possible only along a single "line" between those points. What if, however, time were not linear but more like distance? Then relative times would be understandable. We would say that those who went from one *event* to another would find their times as different from each other as if they had walked different distances between two points in space.

A Quick Look into the Future of This Book

In the chapters ahead, we will look at space and time with new eyes, taking into consideration how both relativity (the science of the very large) and quantum mechanics (the science of the very small) have completely altered what we mean by time and space. We'll look farther into physical time and space and learn why they are considered manifestations of one thing rather than separate categories. We will also explore the notion of sacred time. We will see how time, mind, and spirit have a surprising relation with each other. And we will learn how a mind yoga for time travel springs forth from this relationship, offering surprising benefits and accessible to us all.

CHAPTER ONE

The ANCIENT ART *of* TIME CHEATING

The Supreme Lord said: "Time I am, the great destroyer of the worlds; even without you, will all the people here, all the fighters who took positions on opposite sides, be engaged in destroying.

—Bhagavad Gita 11.32

Yoga practitioners have known about time travel since ancient times, and many still practice it today. Yoga is a system of practice that is part art, part philosophy, and part science. It is a hands-on method for ennobling one's life, finding purpose in it, and going beyond the everyday illusions that inundate us all. According to traditional Indian philosophy, the yoga system is divided into two principal parts—hatha yoga and raja yoga—with many minor divisions within each.[1] Hatha yoga deals principally with physiology, with a view to establishing health and training the mind and body. Raja yoga is a means to control the mind itself by following a rigorous method laid down by adepts long ago. The word *yoga* shows up in several contexts in Hindu thought and has a number of meanings. Yoga is the name of one of the six original systems of Hindu philosophy, which provides the philosophical basis for yoga as presented by the ancient sage Patanjali in the *Yoga Sutras*. In the

Sutras, Patanjali sets forth ashtanga yoga (literally, the eight-limbed practice), which is now generally referred to as raja yoga. Again, the most famous Hindu text, the Bhagavad Gita, talks about karma yoga, bhakti yoga, and jnana yoga—three pathways for attaining enlightenment. The Gita also speaks of kriya yoga, as do the *Yoga Sutras*. When you compare them, you find they complement each other, leading adepts to say that hatha is kriya is raja.

Yoga as both a practice and a system implies a concept of time summed up in the Sanskrit word *samsara.* Samsara signifies conditioned existence, boundedness—the yoking of spirit to spatial and temporal confinement. As Georg Feuerstein, a noted scholar and teacher of yoga philosophy, points out, "Above all . . . Samsara is time."[2] Feuerstein explains that the literal meaning of *samsara* is flowing together—a perpetual flux of things and events producing consequences of causal relationships. As the late Gilda Radner used to remind us on the television show "Saturday Night Live," this flowing together can produce unexpected and undesired consequences—if it isn't one thing or another, "it's always something." This flowing together of things and events has a counterpart in quantum physics, and it is vital to how the mind "creates" time and the appearance of objective events. We will look at this in detail in the upcoming chapters, particularly chapters 8 and 9.

But samsara also refers to something that the Western mind, with its "linear" view of time, does not consider. This is the idea of the wheel of existence—that the soul experiences endless rounds of birth, life, death, and rebirth, set in motion by causal links created in past lives. It turns out that, from a quantum physics point of view, these cycles can be experienced by the time traveler through recognition of the role played by the ego-mind to "anchor" experience—literally bind it into time providing an active focal point or ego.

Samsara is also a term for *maya,* or illusion—the persistent beliefs that bind us to space and time so we participate in the flow of these perpetual cycles rather than escaping from them. This view of life taught by ancient adepts, too, resonates with findings in quantum physics, as we shall see in chapter 8.

Many ancient hymns tell us that time—the past, present, and future—is the progenitor of the cosmos and that time itself is the child of consciousness. Contained within this ancient wisdom is a secret: that it is possible through technique to cheat time—in other words, to travel through time, and even to reach the shores of timelessness. Again, quantum physics agrees, and it tells us how we can draw a map of these shores so the traveler sees what they may look like. I find it striking that modern physics posits the existence of a timeless, spaceless realm of existence without which much of modern physics would make little sense, nor would it connect with reality as we perceive it.

Well and good, you may say, but what does this have to do with time travel? Digging deeper into these ancient texts, we find that they say time and space are products of the mind and do not exist independent of it. The principles of quantum physics, remarkably, tell us the same thing. This is an extraordinary key. The trick to going outside the confines of space and time is to reach beyond their source—the mind itself. Paradoxically, we need a theoretical picture created by the mind to understand what it means to reach beyond the mind. We also need a form of practice.

To make time travel real, not just a theoretical exercise, requires a way of slipping around the corner, peeking under the screen, so to speak, where our usual motion picture of reality is projected. The ancient Vedas referred to this behind-the-scenes look at creation as *kala-vancana,* literally, "time-cheating."[3] It is possible, they say, to escape the space-time illusion of samsara—the projections of the mind itself, which turns out be our own memory in disguise—and cheat time, that is, travel through time. In the coming chapters, we will examine how we think of time and how quantum physics and consciousness are related. But first let's look more closely at what one of the ancient Indian texts has to say.

Chapter One

The Bhagavad Gita

In the early part of the first millennium BCE, Indian philosophers found evidence for the beginnings of what we today call the perennial philosophy. It can be stated in three sentences:

1. An infinite, unchanging reality exists hidden behind the illusion of ceaseless change.
2. This infinite, unchanging reality lies at the core of every being and is the substratum of the personality.
3. Life has one main purpose: to experience this one reality—to discover God while living on earth.

One of the ancient texts in which these principles are set forth and discussed is the Bhagavad Gita.[4] The spiritual wisdom of the Gita is delivered in the midst of the most terrible of all possible human situations: warfare—literally, on the battlefield itself. On the eve of combat, the prince Arjuna loses his nerve and in desperation turns to his charioteer, Krishna, asking him what to do. But Krishna is no ordinary horse-and-cart driver; he is a direct incarnation of God, and he responds to Arjuna in seven hundred stanzas of sublime instruction that includes a divine mystical revelation.[5] He explains to Arjuna the nature of the soul and the nature of the timeless, spaceless, changeless infinite reality and explains that they are not different.

The Gita does not lead the reader from one stage of spiritual development to another, but starts with the conclusion. Krishna says right away that the immortal soul is unchanging and always present and—important for our purpose—that the passing moments of time are illusionary. The soul wears the body as a garment—to be discarded when it becomes worn. Thus the soul travels from body to body, casting aside the old bodies to take on new ones. Just as death is certain for the living, rebirth is certain for the dead. But, Krishna reassures Arjuna, the soul is eternal, not subject to life and death. Arjuna will not be able to perceive this essential truth, however, so long as he remains caught up in life's dualities—samsara, the choices of everyday life in which we are embedded as we move through time.

Like the Buddha's discourses, the Gita does not teach the attainment of an enjoyable life in the hereafter, nor does it offer spiritual or other methods to enhance one's powers in the next life. Krishna's instruction to Arjuna is meant not as an intellectual, philosophical exercise but as a means to arrive at understanding what is truly real. And Krishna teaches detachment as the only way one can get in touch with one's basic spiritual nature. Detachment means not being emotionally entangled with the outcome of our choices. We naturally have the freedom to choose among a range of possible actions in a given moment, but we have no power or say over the results of any act we do.

Is detachment a universally accepted idea? Can quantum physics offer some insight into it? It turns out that detachment is indeed common to both yoga and, as we shall learn in chapter 9, quantum physics. It is profoundly connected to time and the way time works through the mind. The processes of attachment and detachment have much to do with memory and the way we engage with possibilities.

Early in the Gita, Krishna also introduces the word *yoga*—referring not to the physical postures of hatha yoga or to the discipline of raja yoga, but to a certain evenness, or balance, of mind. Krishna encourages Arjuna to establish himself in yoga in this sense, for it leads to profound peace of mind and the ability to be effective in action when it is required. Thus yoga implies acting in freedom rather than through conditioned reflex responses to the events that confront us in life.

Arjuna's Attachment to Time

The Bhagavad Gita contains an episode that each of us in his or her own way finds compelling. Our hero, Arjuna, is facing a difficult situation. About to go into battle against friends who have become enemies, he turned to his chariot driver—who is in fact the divine incarnation Krishna—who has instructed him in the nature of reality and the soul. Krishna identifies himself as the Lord, the source and final outcome of all things, the Eternal One who remains while all living things appear and disappear. Arjuna, always eager to know more, asks Krishna to show himself in this universal divine form.

Lord Krishna then invites Arjuna to peer into Krishna's body and see there at once the hundreds and thousands of his forms, including all things moving and unmoving, and whatever else Arjuna wishes to see: "This universal form can show you whatever you now desire to see and whatever you may want to see in the future. Everything—moving and nonmoving—is here completely, in one place."[6]

The idea of peering into Krishna's formless form is reminiscent of how the mind changes possibilities into actualities, and actualities back into possibilities, from the point of view of quantum physics. Krishna's formless form contains all of his possible forms, just as quantum physics deals with all of the possibilities that matter and energy may assume. Any one form of Krishna appears as an actuality to Arjuna, just as in quantum physics observation by a single mind changes all of the potential forms of matter into a single form. We will explore this phenomenon in chapters 8 and 9.

Arjuna draws a blank. How can he do this? Krishna anticipates his question and declares that Arjuna cannot see this vision with his ordinary sight, so he gives Arjuna the power to see through time. As we will see in the chapters ahead, because of the nature of the mind and consciousness according to quantum physics, the ability to shift from an Arjuna point of view to a Krishna point of view is something each of us already possesses, though most of us do not know this yet.

With this gift installed, Arjuna for the first time sees what he has been missing. But the sight is as bright and powerful as the rising of one hundred thousand suns in a single sunrise. Images appear as if he were looking at thousands of mouths, thousands of pairs of eyes—multiple realities atop multiple realities. Arjuna sees in the single body of Krishna an infinity of images, as if parallel universes were lined up along side of each other like pages in a book.

Figure 1.1. *Krishna shows Arjuna his real nature.*

Arjuna is overwhelmed. He begins to tell Krishna what he sees—the many bodies, eyes, legs, torsos, arms, and so on, with no end in sight. As we see later in the book, this turns out to be a parallel-universe's vision of the sub-spacetime realm, where quantum waves of possibility merge and offer multiple possible pictures of reality.

Then Krishna tells Arjuna the truth: He, Krishna, is Time itself. And with that confession comes a great unveiling: He as Time is both creator and destroyer of all worlds; for, as we know, time favors nothing in its relentless movement. It offers up the beginnings of all things and at the same moment rips asunder all order, all life, all seemingly unchanging forms and, in so doing, begins again to create. The astounding realization is that this power also lies within our grasp. Once we understand how mind and matter interact according to quantum physics, the reason this is so will become clear.

Arjuna's problem—the reason he has turned to Krishna in the first place—is that he is about to go into battle, and the enemies he has to kill are people he knows and loves: his relatives, friends, and teachers. Now, from the viewpoint of this universal vision, Krishna tells him it does not matter whether he slays these people or not: He, as Time, has already killed them. That is, being mortal, they are bound to die. So viewed from this standpoint outside space and time, they are already dead. "From your point of view," Krishna says, "they will all die. From mine, their deaths have already happened. So don't worry about killing them in battle. Take up the duty that has been given to you, and see yourself as my instrument."

After the vision fades from view, Krishna appears in his familiar form again. He tells Arjuna that no one else has seen this complete form of the Lord as the creation, and that such visions are difficult indeed to behold. Krishna explains that people are driven by all kinds of illusions and desires and forget the universal presence of the Lord. But those who are single-minded, free from attachment, and devoted can attain this presence. Only in this way can they enter into the mysteries of Krishna understanding.

What the Story Means for Prospective Time Travelers

What Krishna tells Arjuna at the end reveals an important theme for time travelers. In brief, to time travel you need to leave some baggage behind. Nothing too big—just your ego, is all. For what clouds the time traveler's access to the future and the past is nothing more than the illusion that she is a singular entity, an ego or "I," living in an objectifiable world of time and space. This illusion is extremely difficult to break free of, no doubt. Krishna tells Arjuna that to do this he must become a devotee of Krishna himself. By that he means seeing into the great creator/destroyer that is Time in and of itself. What does this mean in the context of this book? What is the ego, anyway? We will see that it is actually defined in terms of what it does: It acts as a focus for possibilities, a way to change the possible outcomes of a person's life into actualities. In so doing, it also acts as an anchor pinning the mind in time rather than in the timeless realm of Krishna. Just what that means will become clear in the coming chapters.

Krishna reveals something else as well. He tells Arjuna that every soul on earth, whether or not that soul remembers it, desires something that is impossible to manifest. This one desire acts as fuel for all desires. Every soul desires to be one with Krishna, even while remaining in the illusion! That means that each one of us desires to be a supreme master of all that is. From the meekest to the boldest, from those who profess egoless devotion to those who assert themselves controlling dynasties, corporations, or just mastering poses in yoga, all of us are here because we desire to be one with God, the creator, sustainer, and destroyer of all.

Clearly, no individual is capable of taking on the role of the supreme Lord. Krishna realizes this and knows what every soul wants at heart; so to accommodate them all, he gives every sentient form the ability to focus and defocus possibility, which has much to do with the sense that we can control events in our lives. Recognizing also that each one will eventually see the futility inherent in this illusion of control, Krishna nevertheless allows each one to die and reincarnate over and over again, enabling the illusion to persist as long as each being remains enchanted by it.

In this way, each of us may forget Krishna, forget to peer into the body of Time, and enjoy some feeling of power over a piece of the illusionary play. Remembering our desire to identify with Krishna is not easy, even though it is the fundamental desire from which all of our other desires arise.

We each wish to have our cake—that is, to go on living our lives in illusion or *maya*—and to eat it too—that is, realize ourselves as God. Quantum physics may be helpful in explaining why this is difficult to do. When we grasp this, we'll also see why time travel is so difficult.

Space, Time, and Ego

We usually take for granted that the world exists around each of us and that we can experience our own piece of the world through the senses. Physiology tells us how the senses work, how they rely completely on the nervous system to send and receive messages—pulses of electrical activity that travel between neurons. With our modern technological tools, we can witness the tiny electromagnetic signals these messages emit as they move around, particularly in the brain—commonly considered the seat of consciousness.

These electromagnetic "signatures" provide an interesting clue. When we perform certain sensory activities—for example, smell a rose, watch a blinking light, or feel an ice cube on the back of our neck—certain areas of the brain "light up" synchronistically. We now suspect that the patterns of these simultaneous "lightshows" have a great deal to do with our sense of person, in other words, with our ego.

Mysteries abound concerning this thing we call the ego. Up front I want to say that time travel and ego are closely and inversely related, that with the dissolution of ego the possibility of time travel increases. The groundwork for this statement comes from the *Yoga Sutras* of Patanjali, who recognized some 2500 years ago the role our egos play in our lives. In fact, the *Yoga Sutras* can be seen as a guidebook for dissolving the ego.

Patanjali states that there is an innermost part of the individual called the seer, or true self, and a more outer or surface part of the individual called the ego-mind, or illusory self. The ego-mind is composed of three parts: the mind, the intelligence, and the ego. When one learns to master the ways of the ego-mind, pure soul awareness, or knowledge of the true self, becomes possible.

I want to tell you something about the illusory self. The illusory self manifests through five fluctuations or movements of something that exists beyond the confines of space and time—the egoless mind itself. These movements enable the illusion of a coexisting ego and world, or "in here" and "out there," to establish and maintain itself. Consequently, by recognizing these five fluctuations, one can learn to see through them to the true self, much as the child in the famous children's story saw through the emperor's new clothes.

The first fluctuation is perhaps the most difficult to see through because it is so compelling, particularly to a Western view of life. It is termed "valid knowledge"—the kind of knowledge that comes through perception of sensory inputs and learned behavioral patterns. Reasoning and thinking things through is prominent in this movement. Hence logical construction, book learning, classes in school, and the like impress and strengthen this movement. Most of us would feel quite "naked" without our logical minds to hold on to.

The second fluctuation can be called "parody" or invalid knowledge. It is the opinions and beliefs we all carry that are not established by valid knowledge, or the first fluctuation. Herein exist a number of culprits, such as prejudice and hatred, but also romance and fantasy. Many of us wouldn't mind getting rid of hatred, but our love and romance novels—heaven forbid!

The third fluctuation is actually related to the second and consists of "words without meaning or substance," such as exaggerations. Yes, there were overblown opinions even in the days of Patanjali, and in today's world there appear no lack of fantastic claims and over-the-top advertisements.

The fourth movement might not seem to be an obstacle to achieving time travel, but it is. This is the state of sleep without dreaming. "Dreamless sleep" appears to be close to the supreme state of egolessness; nevertheless, it remains an obstacle simply because no devotion or spiritual awareness arises in it. It is the closest one ordinarily comes to letting the self fall away.

There are three states of being: The dreamless state of non-being, the dream state of delusion, and the wakeful state of intelligent awareness. We can see that the dream state correlates with the second fluctuation while the waking state belongs to the first fluctuation. This point implies that dreams are similar to parodies—distortions and illusions. The only difference is that they occur to the sleeping mind instead of the awake mind.

The fifth fluctuation is most likely the hardest to deal with in practice. It is the memory of all perceptions, imaginations, thoughts, objects, senses, and interactions with others. The thing of it is, memories persist because we have made an effort to hang onto them. A long time ago, even as a kid, I realized that I could not memorize anything unless it had some emotional effect on me. That is very nice when you are trying to remember your first kiss or that funny joke your uncle told you, but it has a downside. Once you do have a memory, it remains persistently fixed in the mind—even if you think you can't remember it. It lurks there in your unconscious, ready to jump out at you the first chance it gets. It doesn't take much to recall a hidden memory, unless it's one that the deep mind decides to keep hidden. A dream will do it, as will meeting someone who reminds you of someone you met long ago.

You can think of these fluctuations as barriers—impedances placed in your way. They plant you firmly on earth, in ordinary time and space. They act as traditional anchors, and with them well in place, time travel is virtually impossible using your mind. Yoga schools offer the means to go beyond these barriers. They define the goal as realization of the true self, "beyond" the fluctuations that keep the illusory self alive and well. In the third chapter of the *Yoga Sutras*, Patanjali describes time travel—knowledge of the past and the future—as an ability or "power" that is possible along the way to this realization.

CHAPTER TWO

2

A BRIEF OVERVIEW *of* SACRED TIME *and* SPACE

The perception of duration itself
presupposes a duration of perception.
—Edmond Husserl

To realize the "true self" is a task that may not be easy for a number of reasons. Why should it be so difficult? One cause is that we live "in" space and time. This answer is easy to articulate but hard to appreciate fully. The problem has to do with the reality that, while time and space seem to be "out there" as objective facts, they also turn out to be deeply ingrained in the "in here" world of the mind. We can think of the "out there" world as ordinary or profane and of the "in here" world—although often chiefly concerned with objective events—as a sacred stream of time at its very core. Sometimes this sacred stream does not run at the same "speed" as the clock-on-the-wall ticks.

University of Texas Professor E. C. G. Sudarshan tells the following mythological story from the Vishnu Purana that illustrates this connection.

> In the Vishnu Purana there is a mythological story about sage Nárada asking Lord Vishnu to tell why people are deluded into living in profane time when all along they could function in

> sacred time. Lord Vishnu offers to do so, but asks Nárada, in the meantime, to fetch a cup of water. Nárada goes to the nearest house and knocks on the door to ask for the water. A beautiful and attractive young woman opens the door. Nárada is completely captivated by her charms, forgets about his fetching a cup of water for the Lord, forgets that he is an avowed celibate; and he woos and wins her hand. They live together in a house after getting married and in due course two beautiful children arrive in successive years. While they are living in contentment, suddenly a flash flood engulfs their neighborhood and even their home. They have to try to escape as the flood waters rise and the current becomes stronger. It becomes so strong that first one child, then the other, and finally his wife are swept away by the raging waters. Nárada himself is barely able to maintain a precarious hold on a tree and is feeling terribly shocked by the tragedy that has befallen him. While waiting thus, he hears Lord Vishnu's call asking him "where is the cup of water" because he is still thirsty. Nárada suddenly realizes that he was all the while standing on the firm ground and only a few moments had passed![1]
>
> —E. C. G. Sudarshan

Most of us have experienced, at one time or another, the distinct feeling that time has passed too quickly or perhaps too slowly. I know that when I sit down to write a book such as this one, I struggle for several minutes at the beginning, but once I find a rhythm and the words begin to flow, I lose all sense of time. Perhaps hours go by and I have no sense of their passing at all. On the other hand, time seems to go much too slowly if I find myself in an embarrassing situation or when I'm visiting the dentist and experiencing the dentist's drill. Scientists, particularly psychologists, call this relative experience of time "subjective time."

Objective time, by contrast, is that "thing" we believe to be measurable by clocks and in terms of rhythms or frequencies. In fact, all clocks work by comparing rhythms—they imply an objective time simply by counting repetitions. Now this may not seem to be a comparison of rhythms, but it is most certainly that. For instance, if you choose to count the number of swings of a pendulum, as Galileo did one morning long ago in a Sunday service watching a swinging chandelier, you are actually comparing the number of swings you see with your own subjective internal rhythm—for example, your heart rate or your eye blink rate or even the rate at which words arise in your mind. Think about it: How do we know that a pendulum makes a "good" clock—one that keeps "true" time—except by comparison? (Note how the assessments "good" and "true" subtly enter the picture here.) Certainly we do compare a questionable clock with another that we trust keeps good time. Yet even though we may check our clock with a trusted timepiece, we perhaps most often notice that our mechanical clocks are incorrect through comparison with our inner time sense.

The human mind is capable of discerning the differences among a vast array of rhythms—from the amazingly rapid vibrations of the quartz crystal in a watch to the yearly journey of the earth around the sun—and, based on those differences, constructing an objective "timescape," a vista or expanse of time that all of us see and agree on. To make these comparisons requires an internal, subjective sense of time.

However, as we saw when we examined the five fluctuations of the mind in chapter 1, this time sense may be an illusion causing us to think that something that *has* happened is happening now, or will happen again. This inner, perhaps illusionary connection given to us by the great God of Time turns out to be the first tether that binds us in time and space and subjects us to time. Without this connection, the vibrations of music and sound could not play a vital role in enchanting us, nor could the sun's rising, the movement of tides, and the changing seasons. Yet despite the fact that these natural rhythms are cyclical, we in the West have interpreted them to mean something quite different. We have learned to map them linearly, implying that even though they repeat, they never repeat themselves in quite the same way. What is it that is changing? This sense that something changes gives us an experience we label "time passing," and we have learned to see that experience in terms of a straight line.

Chapter Two

A Line of Time

The notion of linear time is an objective construction of the human mind, one that is particularly ingrained in the Western attitude toward life. We in the West give more credence to objective, or mechanical, clock time than we do to our inner, subjective time sense. We ultimately reduce all subjective senses of time to the merest thread of objective agreement. Yet our inner, subjective sense of time is as real as any sense can be. We think that since we can't measure it, it can't be real. But what could be more real to us than the inner sense of time through which we experience rhythmic variations like music and even the pace of our own thoughts and feelings? We may not be able to compare it with another person's temporal sense, but this shouldn't make it any less real.

We have abandoned our inner sense of time, not because of the Gita's teaching, but to replace it with the commonly accepted outer sense we call clock time. Yet linear clock time doesn't really exist "out there" any more than subjective time does. It, too, is abstract and imaginal. But based on that imagined, objective thread or line of time, we produce an enormous outflow of creative and technological innovation. We construct, for example, the notions of the forty-hour work week, the nine-to-five office, the daily grind, the two- or three-week vacation, equal employment opportunity, equal hours of work for all employees, overtime, slacking, and so on. As for technological inventions, nearly every one of them implies linear time at its heart. For what are inventions but devices to save time so that we can increase our hourly, daily, and yearly output—or else to help us pass the time that we've saved?

We walk on a temporal tightrope that stretches from the instant of our birth to the last breath we take. This linear notion of time appears to make sense to us, and it certainly seems egalitarian and "real"; nevertheless, it arises ultimately from a subjective perception. Inside our minds lies a sense of time that tells us, even without a watch on our wrists, what takes a long time and what doesn't. We hone this sense of time as we perform any number of daily tasks, from waiting in line at the grocery checkout stand to brushing our teeth before we retire. Clocks and calendars certainly were invented to display this inner sense of time, allowing us to make comparisons. For without comparing clock time with our inner, subjective sense of time, we would have no measure of the difference between our dreams and fantasies and the reality we presently believe we are living in.

Without this inner temporal sense, we would not be able to measure the length of a thumb or the height of a tree or—for more sophisticated examples—the height of a skyscraper, the flying altitude of a modern jetliner, or the distance to the sun and other stars and galaxies. Our inner temporal sense enables us to realize and measure space, simply because it takes time and repetition to do so. It may not seem that you are repeating anything when you use your eyes to measure the length of your thumb with a tape measure, but the light reaching your eyes consists of many frequencies, and these rapid repetitions in turn provide you with a sense of sight.

Many other Western societies have also developed the line-time idea. In fact, one way or another, at times with some difficulty, all civilizations have adopted or formed a concept of linear time—one that shaped their attitudes and enabled them to have a historical perspective and anticipate the future. Professor Sudarshan reminds us that the two great civilizations of Asia, the Chinese and the Indian, have treated time differently from the way Western civilization does. The Chinese kept meticulous chronology, but valued ancestral time more than present time. Immediate ancestors were held in highest regard, and the duty of the individual was to do hard work for the good of society. As long as the people worked hard and kept the ancestors in mind, society would progress and life would be better for all. Indian society, on the other hand, "seems to have the notion that time as experienced depends on the state of awareness of the individual, and hence time functions in a variety of subjective forms. So chronology in India is unreliable, in any linear objective sense, and most events were simply 'a long time ago.'"[2] That is, the Indian mind does not see time as a simple imaginary scaffolding—something projected by the mind "out there" as a skeleton or framework upon which the real business of the world is measured and compared. Instead, time exists integrally and inseparably from space and matter; as a result, it can change in a nonlinear manner.

Cycles and Dreamtime

The Chinese and the Indians aren't the only peoples who look at time differently from the way Westerners do. In a chapter of my book *The Dreaming Universe,*[3] I write about the ways of the Australian aboriginal peoples. In his book, *White Man Got No Dreaming,*[4] W. H. Stanner refers to the Dreamtime, or the *Alcheringa,* of the Arunta or Aranda tribe, first introduced to the West by two Englishmen: anthropologist Baldwin Spencer and researcher Frank Gillen.[5] Stanner prefers to call it "the Dreaming" or simply "Dreaming." "Dreamtime" is a curious term. Surprisingly, it is not original to the Australian aboriginal people. Rather, it was coined by Gillen in 1896 after his attempt to understand the aboriginal concept of time and was used by Gillen and Spencer in their now-classic work of 1899.[6] Even though aborigines think of Alcheringa not so much as Dreamtime but more as the law or the sacred understanding of life, time nevertheless enters into it.

The Dreamtime refers primarily to a time of heroes who lived before nature and humans came to be as they are now. It was a time long ago, as in "Once upon a time, there was. . . ." That is, neither time nor history are actually implied in the meaning of Dreaming. Time as an abstract, objective concept does not exist in the aboriginal languages. The Dreaming cannot be understood in terms of history, either. The Dreaming refers to a complex state that eludes the Western linear description of time and Western logical ways of thinking.

According to Australian scholar W. Love, early Australian aboriginal people, when they arrived in Australia sometime between 40,000 and 120,000 years ago, were faced with flora and fauna very different from what they had known in their own land.[7] These macro-fauna, as Love calls them, became in myth and legend the animals of Dreamtime, and their stories became models for human behavior and were enshrined in ceremonial patterns. As Stanner explains, an aborigine may regard his totem, or the place from which his spirit came, as his Dreaming. He may also regard tribal law as his Dreaming.

According to another expert, Ebenezer A. Adejumo, Dreamtime was not just a fantasy of aboriginal people.[8] Instead, it has as much meaning to them as psychologists and psychiatrists place in our dreams of today. The myths of the Dreamtime contain records associated with certain geographic sites, sociological concerns, and personal experiences. Since the aborigines reenact the stories of the Dreamtime through ritual, we can deduce that all of the past, present, and future coexist in the Dreamtime as if in parallel worlds of experience. Together these realms make up a reality in which our sense of present time is merely a small part.

The Dreamtime is eternal and timeless, and so are the spirits of the people who are linked with it: They have existed in the past, they will exist in the future in the hearts and minds of the children yet unborn, and they exist now in the hearts and minds of the people of the land. Aboriginals see both themselves and all human beings this way. There is no division between time and eternity; all time is essentially present time. To keep this awareness alive, songs must be sung, dances must be performed, and these creative acts become the repeated reincarnation of the spirit reenacted by countless repetition by human forms. By keeping track of the stories and legends, the spirit is in a real sense keeping track of himself—his path and pattern throughout historical time.

This reenactment serves as a solution to the alienation of humans from their own planet. We are all utterly dependent on the earth for survival. The aboriginal culture does not view nature separately as our Western scientific world does, thereby adjusting itself to life on earth through applied science. Instead, it sees itself as part of nature.

Australian aboriginal people today are well versed in linear time, yet they still refer to time in their own original manner. Hence their grammatical constructions in English may seem quaint to Western ears, but I assure you, their use of English is quite correct in terms of their own sense of time. As in a poem one old black "fella" once told Stanner:

> White man got no dreaming.
> Him go "nother way."
> White man, him go different.
> Him got road belong himself.

Time for the aboriginal is quite concrete. It is based on the observance of natural rhythms, such as the seasons and the lunar and solar cycles. Thus time is marked, not by points on a line stretching from minus to plus infinity, as in the Newtonian worldview, but on a circle: Time is counted by recurrences of cycles. The timing of daily events is marked by the position of the sun. Natives of central Australia mark time in "sleeps"; they say they will return to a place after so many sleeps, or nights. Durations of time are marked by everyday processes. For example, one hour may be marked by how long it takes to cook a yam. A moment might be the twinkling of a crab's eye. Longer times may be marked by the duration of a particular journey. Thus time tables are not definite. What is important is the concrete time of the "now."

When time is viewed as circular and sacred, it appears to have an imaginal quality. This imaginal quality is not unique to the aborigines. I believe all humans sense the imaginal quality of time. But we in the West tend to dismiss this subjective perception of time in our commitment to a line-time view of events. I like to think of time's imaginal quality as a great hoop that rolls along the imagined straight line of our linear time. I'll say more about this cyclical sense of time in chapter 4.

CHAPTER 3 THREE

An OVERVIEW *of* PHYSICAL TIME *and* SPACE

Time is that quality of nature which keeps events from happening all at once. Lately it doesn't seem to be working.

—Anonymous

Before we can consider time travel as a reality, we need to examine time itself, to make sure that when we talk about it the metaphors we use—and we do need metaphors to talk about time—don't get us into quandaries from which there is no extraction. A good place to begin is with what some past masters of time, both ancient and recent, had to say. Interestingly enough, they all recognized that time and mind are not as separable, and time is not as objective, as we might believe. In fact, they suggested that time is projected from mind in some manner.

Saint Augustine, who lived in the late fourth and early fifth century, wrote: "What, then, is time? If no one asks me, I know what it is. If I wish to explain it to him who asks me, I do not know." Perhaps Augustine was referring to how difficult it is to reconcile our common, subjective sense of time with our objective, mathematical description of time. Or perhaps the saint from Hippo was jumping into the future and talking about that peculiar quantum physics paradox called the complementarity principle, which says that (from a psychological point of view) you cannot express your knowledge without altering it. By that principle it is probably safe to say that no one can really articulate what time is simply because we have nothing really to compare it with. Although we certainly try to find proper metaphors, it seems that time doesn't have any objective qualities.

Even though we don't quite know what we speak of when we talk of time, we nevertheless spend a lot of our time doing so! Who doesn't carry a wrist watch these days or, if not, find oneself asking what time it is or looking for a clock? The best we can do when dealing with time is imagine that we can grasp it, and in that imagining we form a model so we can think about it. Mathematical language may even affirm that model and provide a picture of what we are talking about. When we speak of time travel and time machines, it's good to acknowledge that although what we're talking about may seem perfectly obvious, in fact we're not really very sure about it.

Let's return to wise old Augustine. One wonders if he in some way realized the theory of relativity, for he goes on to say: "Yet I say with confidence that I know that if nothing passed away, there would be no past time; and if nothing were still coming, there would be no future time; and if there were nothing at all, there would be no present time."[1]

Compare this with two modern-day statements on the mystical or mindlike nature of time: "People like us who believe in physics know that the distinction between past, present, and future is only a stubborn persistent illusion."[2] "What we mean by 'right now' is a mysterious thing which we cannot define. . . . 'Now' is an idea or concept of our mind; it is not something that is really definable physically at the moment."[3]

The first quote is from Albert Einstein and the second from Richard P. Feynman—both, of course, Nobel Prize-winning physicists who had much to say and offer on the subject of time. Both Einstein and Augustine refer to something we feel inside ourselves. Augustine tells us that without these feelings of something passing away, something within our midst, or something about to occur, we would have no awareness of the past, present, or future. Einstein tell us that these feelings are not true about time *as it is* but refer to a trick of the mind—an illusion.

Certainly all of us have experienced at some point what I call a timeless moment—a period of quiet reflection when everything seems at peace. The reality of this moment surpasses the illusion of time. Feynman tells us that what we mean by the immediate present is mysterious and undefinable—whatever we say about this mystery remains a projection of the mind, and that projection has no physical existence, hence there is no way to define or measure it.

The Physics of It All

Of all the subjects studied in schools and universities these days, physics is rarely the most popular. I often wonder why. I admit that my high school physics class was less than I'd hoped for; it was sometimes quite boring. But by the time I reached college, physics had become the center of my interests. I was always curious about light, particularly how it moved and what it was made of, and I was fascinated by color. Although I didn't find the lab work nearly as exciting as the ideas of physics, setting up optical benches and spectrographic equipment and measuring some of light's many mysterious properties had me completely fascinated.

The nature of light has been at the core of nearly every revolution in our thinking about the universe. Light comes in many forms: not only the small portion of the spectrum our eyes can sense, but also ranges of the spectrum above and below it—radio waves, x-rays, infrared rays, microwaves, and so on. For our present inquiry into the universe concerning time and time travel, two qualities of light are particularly relevant. The first has to do with how fast light travels. The second has to do with the way it travels to our eyes or our measuring instrument.

The speed of light is not easy to determine. The most refined measurements tell us that in empty space, light travels at a constant speed, somewhere around 671 million miles an hour. When you are moving either toward the source of light or away from it, you might expect that you and your measuring instrument would record slightly different speeds for light, depending on how fast you are moving, even though your own speed is so significantly smaller than lightspeed. However, instruments refined enough to detect these small differences have never been able to do so. How can it be that the speed of light doesn't change when you move relative to the source of that light? The answer to this puzzle was discovered by Einstein, who theorized that it wasn't lightspeed that would change. Instead, it's the assumptions we use when we measure space and time that had to change. He was not referring here to the clocks and rulers we use to measure time. Rather, he meant that our very concept of what we are measuring has to shift.

Just how does light travel? Does it rush from one place to another as individual objects, like tiny particles or bullets in a vacuum? This is the way Sir Isaac Newton imagined light in the late 1600s. Or does it undulate along like waves in water, as Thomas Young proposed nearly two hundred years later? Based on experiments he carried out around 1820, Young announced that light moves, not like the bullets that Newton had proposed, but more like waves in water. Hence the paradox called the wave-particle duality was born, and it remains a mystery even today for those who believe light must be one or the other.

How and Why Time Flies Slowly

In the introduction, I explained how tiny particles called muons normally live very short lifetimes, but when they are given the opportunity to move quickly enough—near lightspeed—they exhibit the strange property of living longer than they should, as much as ten times their normal lifespan. This discovery about these minute particles has far-reaching consequences for our understanding of time and even for our understanding of Einstein's special and general theories of relativity.

To say that time is relative may sound like mere erudition, but what it actually means is quite amazing. According to both relativity theories, the time experienced by any thing (such as a clock or a person) depends on how that thing moves relative to anything else; if it moves in the most effortless and natural pathway possible, the time will be shorter as compared with the time experienced by other, nonmoving things. This discovery is known as the *time dilation effect.* Sounds tricky? After we have examined how this works, we shall be in a good position to grasp just what time travel really means and why I would even consider mind yoga as the means for time travel.

Physicists are very interested in discovering the rules of the universe, how these rules got formulated, and, if possible, how they may be broken, or even how new rules can be discovered. Such is the work of physics—nothing less than seeking to understand God's handiwork at the deepest levels. Every once in a while, a principle of great depth is revealed and physics is immensely strengthened by it. One such principle has been called various names but usually goes by *the principle of least action.* In essence it says that things move in whatever way makes the universe more efficient—as if God does not wish to waste resources, even while She or He wishes to make things as interesting and varied as possible.

Einstein recognized this efficiency principle and incorporated it into his general theory of relativity. In physics lingo, this principle is stated slightly differently. Instead of *least,* the word is *extremum,* which means the absolute minimum *or* the absolute maximum value that anything can have. In spite of the complexity of the universe, when things move naturally—no matter how or where they go—they tend to move along *extrema.* In other words, something about their motion takes on either a maximum or a minimum value.

For example, imagine yourself sitting still in a spaceship traveling to Mars. Your path in space is natural, or effortless, enough, and you are moving along at a constant speed on an extremum through space. If the astronaut sitting next to you in the spaceship were to move away from you in another spaceship and then return and the two of you compared clocks, his clock would show *less* time had passed than yours would. On the other hand, if you were to compare your clock time with that of another clock that was at rest compared to you, your time would be less. The gen-eral rule indicates that time itself depends on motion and specifically how that motion takes place.

According to both the special and the general theory of relativity,[4] the time you spend moving along any naturally occurring trajectory through space turns out to be an extremum when compared to the time observed by another person who also started out when and where you began and ended where and when you ended but followed a different trajectory. The starting and ending points for two or more such observers are called *events.* It would seem that since they both started together and ended together they would both pass the same amount of time. But in relativity appearances are deceiving.

In figure 3.1, we see two events, A and B, separated in space and time, and two paths or trajectories beginning at event A and ending at event B. Even though both start from the same event and end at the same event, the trajectories are very different. A clock that traveled along the curved path would actually record less time passing than a clock that traveled along the straight path.[5]

It turns out that the general theory of relativity and many other laws in physics are derivable from looking at such paths through space and time and noticing which of the paths maximizes the amount of time spent in following the trajectory. Remember, we are only comparing clocks that move along different trajectories but have the same end points. If we compare the time spent on any one of these trajectories with the time indicated by clocks that are not moving, the moving clock will always show less time. Among those moving clocks, one of them will

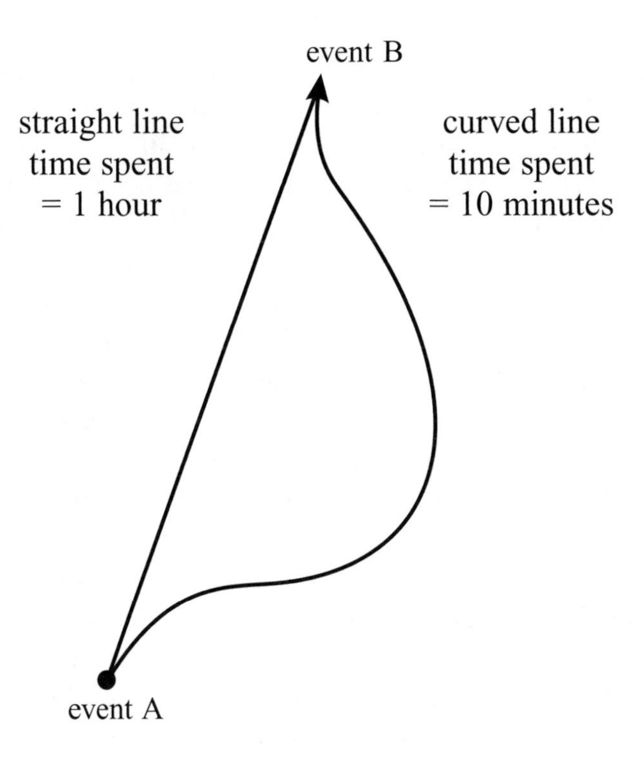

Figure 3.1. *Paradoxically, in spacetime, a curved line takes less time than a straight line.*

show the most time passed. The trajectory traveled by that clock is called a *geodesic* in the language of the general theory of relativity. In figure 3.1, the straight path is a geodesic while the curved path is not. This word *geodesic* actually means the shortest distance between two points on our planet. Physicists working with relativity theories borrowed the word to mean the extremum between any two events separated in space and time.

Let me give you some simple spatial geodesic examples. You have heard that a straight line is the shortest distance between any two points. This is true on a flat surface. If you draw any other line, such as an arc, between two points on a piece of paper, the length of the arc will certainly be longer than the straight line. Hence the straight line is an extremum—in this case a *minimum* length extremum. However, a straight line is not an extremum on all surfaces. In fact, extrema depend on the shapes or curvature of the surfaces in which possible trajectories are drawn. Take a sphere in hand, if you have one—if you don't, an apple or orange will do. You can't draw a straight line on a sphere, can you? Your line must follow a curve. The shortest-length curve you can draw between any two points on the sphere forms a segment of a *great circle,* which means any circle on the sphere with the *maximum* circumference possible. Segments of such arcs form minimum-length extrema.

Indeed, in this age of transcontinental flights, especially transoceanic flights, airplanes follow great-circle routes to go economically from one place to another. It's a bit surprising to find we must fly northward and then southward to reach London from Los Angeles. We are used to thinking of the earth as a flat surface—a habit that comes from looking at planar maps wherein Delhi is west of San Francisco, while Moscow, for example, is considered to be to the east. Europeans refer to countries like India and China as the East, or the Orient (a word that means "east") because it is traditional to think from a flat-earth, European point of view.

Hence a straight line between Los Angeles and London drawn on a flat map will not even be close to the actual shortest length between those two cities; a great-circle route that heads northward, nearly reaching the pole, and then southward across the Scottish Highlands into London will be the shortest route. Great circles are examples of spatial extrema, usually the shortest distances between points.

However, when we begin to consider time along with space, as we do in the general theory of relativity, the extrema change. It is quite a surprise to find that extrema of trajectories that consider both space and time are those that take the *longest* time periods when compared with other trajectories between the same endpoints. Let me take a little more "time" to explain this.

Time and Space Are Inseparable

Modern physics has shown us that time and space are not separate and that our perception of them as separate is an illusion. To be able to treat them as inseparable, physicists came up with the term *spacetime.* We owe this new perception of things to Albert Einstein's famous 1905 proposal, later named the "special theory of relativity." I'd like to examine this theory—not the physics so much as what it means metaphysically and philosophically—for the wonderful insights it offers regarding time and time travel.

We may not be able to appreciate the full impact of Einstein's proposal today, for during the past fifty years or so we have become more receptive to nonmechanical and even spiritual ideas making their way into scientific thinking. But in the late nineteenth century and beginning of the twentieth, Western society had quite a mechanical view of the world. A rigid materialism had developed and spread across Europe, and soon after, along with the tide of European immigrants, the currents of culture had carried it across the Atlantic to the United States. A quiet kind of arrogance ensued as many scientists claimed the age of scientific discovery was over.[6]

Einstein's seminal 1905 paper, "On the Electrodynamics of Moving Bodies," upset the reigning scientific belief systems of the day. The popular press struggled in vain to report on it, for this radically new idea, couched in mathematical formulas, simply made no "common" sense. Einstein had unlocked the riddle, not just of an observable problem having to do with the relative motion of wires carrying electric currents to electric fields and magnetic fields, but of the very nature of human observation of physical facts. His equations traced out how it was that two observers, moving relatively to each other in a smooth and unaccelerated manner, could come to different conclusions about the when and where of an event they both observed. (In case you are wondering what "moving relatively" to each other means, think of two rocketships, each moving at constant but different speeds. It doesn't matter in which direction they are going relative to each other; all that matters is that when each one sees the other, the other will appear to be moving farther away or closer.)

Before 1905, it had been firmly believed—and it made the most "common" sense—that time and space were separate and each of them was immutable. Hence, regardless of how the two observers moved in relation to each other, they should always agree on where and when an event took place, assuming they each had an accurate yardstick and clock.

The problem was how to compare measurements when two observers were far apart. To do so meant using some form of signaling, and if the two observers were really far from each other, light (or radio waves, which move like light) was really the only means available. The question was: How did light move from one place to another, and would it appear to move at different speeds relative to different moving observers? Around 1887, Albert Michelson and Edward Morley set out to measure the speed of light. The test required refined measurements of light moving in different directions. Since Young's experiment had shown that light behaves like a wave, the question was, in what medium was the lightwave waving? Even though everyone thought that space was empty, through it light must travel from the sun to the earth. Thus light must be waving through something as it makes the trip. This seemingly invisible medium was called the ether. Like a great ocean in space, the ether was thought to fill all of space, and everything that moved had to move through it. Since the earth moved in a large, nearly circular ellipse around the sun, it had to move at first in one direction relative to the ether and then in a different direction months later. This should result in the presence of an ether wind affecting lightspeed on earth, just as a wind effects the motion of any object moving into it or with it. Even though light moved at the extremely high speed of 671 million miles an hour and the earth moved much slower than that around the sun, Michelson and Morley attempted to measure the difference in the speed of light due to the earth's motion relative to that ether. To their surprise, they found no difference.

Einstein showed that even though it made no common sense that two observers moving at different speeds would see light coming from a common source as traveling at the same speed, nevertheless, it had to be true. And to accept this truth about light's speed meant giving up our ordinary notions about space and time. Scientists had as much trouble grasping this as did the general public.

For example, consider our sun, around eight lightminutes (about 93 million miles) away, shining light in its usual manner.[7] Suppose that another observer who is moving very quickly relative to us—perhaps at nearly lightspeed—is also observing our star. Suppose the sun suddenly goes a bit berserk and emits a gas cloud that temporarily occludes the light (a so-called sunspot). Then a few minutes later the sun calms down and the "spot" disappears. For us, these events would appear to occur at nearly the same place in space (perhaps slightly shifted due to the earth's rotation), and within a few minutes of each other. But the other observer would have an entirely different experience. She might see the two events separated by years of time and millions of miles in distance.[8]

Between 1905 and 1908, all of the mind-boggling examples arising from seeming paradoxes in relativity remained just that—no one could really understand them rationally or even intuitively, in spite of the mathematical success and the logic behind that math. Einstein had certainly shown that we couldn't think of space and time as we were accustomed, but the notion of spacetime as a single entity hadn't really occurred to him. All this began to change in 1908, when Albert Einstein's former teacher Hermann Minkowski presented a popular lecture to a forum of German scientists and physicians in Cologne.[9] His lecture, "On

Space and Time," was based on a paper he wrote entitled "Basic Equations for the Electromagnetic Phenomena in Moving Bodies," which was based on Einstein's earlier work. Minkowski's lecture heralded a new era of "mystical physics"—one that has opened the door to such topics as time travel. I believe it provided the necessary basis for altering Western culture, making the age of information and quantum physics possible. Minkowski said, "Gentlemen, the ideas on space and time I wish to develop before you grew from the soil of experimental physics. Therein lies their strength. Their tendency is radical. From now on, space by itself and time by itself must sink into the shadows while only a union of the two preserves independence."[10] By "independence," Minkowski was referring to the idea that only this union did not depend on anything else, such as the amount of matter or energy present. However, when Einstein introduced his general theory of relativity in 1916, even this independence was shown to be a fiction.

That time and space are not independent of each other can be difficult to grasp. Space seems to be infinitely "out there," filling a huge blackness, an enormous emptiness, a great void, while time seems so fleetingly "in here," personal, subjective, and hardly ever brought to our attention unless we are watching a clock or timing events. When the first Mars landing took place on July 4, 1997, it was amazing to learn that commands issued from the Jet Propulsion Laboratory in Pasadena, California took more than ten minutes to reach the tiny Mars rover ambling across the great Martian plain, 111 million miles away. These commands were carried by radio waves—just another form of light energy—and hence were limited in their ability to alter the rover's movements by how fast the signals could travel. You may have had the opportunity, as I did, to view the live television coverage of the rover's first steps. I remember how eerie it felt, watching the rover and realizing that the picture I was seeing on the screen had had to travel for more than ten minutes at the speed of light to reach me. It drove home the fact that space and time are absolutely connected, a single thing we can call *spacetime.*

The Notion of Proper Time

As physicists accepted that space and time are inseparable and began applying the general and special theories of relativity, they began to wonder what to call the time experienced by a clock traveling along a spacetime trajectory when compared with the time shown by other clocks that are not traveling on the trajectory. They decided to call the traveling clock's time *proper time.* No, proper time isn't the polite hour you spend visiting your rich great aunt Mary in her mansion at high tea; rather, it refers to the time you spend moving along any trajectory in spacetime from a beginning event to an ending event. Other travelers could have followed different spacetime trajectories between the same two events and their clocks would show different proper times. One of these trajectories would be the geodesic connecting the events, and it would show the maximum amount of time possible between the two events. However, I remind the reader that what is a maximum and what is not depends on who is doing the measuring. Nonmoving clocks will always show more time lapsed than moving clocks. Moving along a geodesic means moving along the most natural or effortless pathway and the longest time possible (but always a shorter time than other nonmoving clocks indicate) between two events. However, as we have seen, this is true only on a flat plane; on the surface of a sphere, the geodesic turns out to be a curved line through three-dimensional space.

Let me summarize: Proper time is a measure of the time spent along a spacetime trajectory as measured by a clock moving along the trajectory. Proper time is always shorter than nonproper time. In other words, moving clocks go slowly when compared with clocks at rest. These other clocks, usually at a distance, could measure the time spent by an object moving along a trajectory, but they would not be the proper time of the object. Spacetime geodesics mark the greatest proper time between events when compared with any other trajectories having the same endpoints.

As an example, suppose you and your brother are sitting together on the living room couch. You just sit there without moving from 1 P.M. Tuesday to 1 P.M. Wednesday, the epitome of a couch potato. However, even though you don't move an inch in space, you are moving along a geodesic in spacetime, since the earth is turning and also progressing along its orbit around the sun in a natural way. The amount of time passed that you observe on your watch is a proper time of twenty-four hours. Meanwhile, your brother gets up from the couch at 1 P.M. Tuesday, rushes out the door and climbs aboard a jet airplane. He flies at 600 mph for eleven hours and then gets on the next plane back, returning to your city in about eleven hours. Give or take traffic conditions, he makes it back to the couch at 1 P.M. Wednesday, as witnessed by you. After he arrives back home, you sit down together and compare watches. Even though he sped through the universe for a while, as far as he was concerned he was at rest in his airplane, with his watch showing his proper time. Since he had to artificially accelerate when taking off, landing, driving to the airport, and so on, he figures his trajectory was not a geodesic when compared to yours. Was he correct?

You compare your watch to his, assuming you both have very high-resolution atomic timepieces, accurate to a few nanoseconds (a nanosecond is a billioneth of a second). What you will find is that his watch shows less time passed than yours—not much, just about 35 nanoseconds less. If he had traveled at, say, satellite speed, around 25,000 mph, that time dilation would increase to around 60 microseconds, or 60,000 nanoseconds. If he had been able to move at lightspeed, that dilation would come to twenty-four hours precisely—meaning that he would not age one nanosecond during the twenty-four-hour period you sat lethargically on the couch aging a full day. That's right: Moving at lightspeed, even though the time spent is all that can be spent, turns out to be no time at all.

So your brother *was* correct that his trajectory was not a geodesic. Both the traveler and the nontraveler experienced proper, but different, times between the same events. However, since the traveler moved along from one point to another and then returned, his path through spacetime was not a geodesic; hence, he experienced less time than the nontraveler.

Time *and* Space Enough?

Time is an illusion.
Lunchtime doubly so.
—Douglas Adams

Just how big is space and how long is time? When writing or speaking about space and time, we find ourselves in a quandary of difficulties. The problem has to do with language itself, for the words we use are mainly metaphors arising from our experience, which of course is embedded in space and time. Metaphors allow us to compare one experience with another. Through such a comparison we come to feel we understand the new experience in terms of the old. But often experiences arise that simply don't fit the old metaphorical description.

In this chapter, we will examine some of the more remarkable metaphors and models people around the world have used to describe and think about time and space and consider the possibility of stepping beyond these limitations. The question then arises: Are we really transcending the boundaries of space and time, or have we simply freed ourselves from the subtle traps laid for us by our own languages?

For example, in English, or in any Indo-European language, we use past, present, and future tenses, assuming that temporal experiences can always be mapped this way as if they were points on a line. If something happened yesterday, it *was* in the past. If it happens tomorrow, it *will be* in the future. When we think about and relate an event, when and where is usually important. But not all peoples consider experience this way. Some languages are based on descriptions of the feelings one has during an experience rather than when and where it occurred. In these languages (Native American languages such as the Navaho are an example), distinction between past, present, and future are not necessarily referred to as significant.

But in Western Euro-American culture, spacetime comparisons are important. We ask, "How long did it last?" when we talk about an event. In fact, "long" refers primarily to length—a measure of space, not time. So what do we mean when we talk about how long an event lasts? We also, in our impatience, want to know how long it will take us to make a change in ourselves, to become adept at some task or practice. In the upcoming chapters, we will see how time travel through mind yoga can be used to slow down and temper impatience, as well to speed up and move through events quickly when needed.

Experiencing beyond Spacetime

Is it possible to step outside the limits of spacetime? If spacetime is a metaphor, or if it is unreal, it must be possible. Ancient wisdom suggests that neither time nor space is real, nor are they separable from the ego, or I-sense. Ancient texts ask us to explore these questions and answers: If we are nothing but our bodies, and the world we perceive is nothing more than what our body-mind complex tells us, then it seems we are embedded in time and space. But is this the case? Our senses may try to convince us that we are the same as our bodies at all times and in all places. But if time and space do not exist apart from the sense of "I," then, the sages tell us, we are a part of some magnificent reality that transcends time and space.

How can time and space be projections of the mind, coming into existence subjectively along with the ego—a notion that flies in the face of our experience of them as objective realities embedded "out there," independent of mind? Consider the example of deep sleep, or dreamless sleep—a state that dream researchers say we pass into before we begin to dream. In deep sleep the ego disappears; we have no sense of who we are. The ancient texts point out that when a king is asleep, he forgets he is a king; when a beggar sleeps, he forgets he is a beggar. In this state there is no experience of time or space; the ego is not present to create it. When we dream, we do experience a dream landscape composed of time and space. We also experience a dream ego—the one who is experiencing the dream. The dream seems real to us when we are in it. But it's clear to us on waking that the dream landscape was created by our minds. When we awake, we also sense an ego, this time identifying with our physical body, experiencing "I am the body." We also experience time and space and locate our body in it. But who is to say that the objective, waking reality we experience daily is not a form of collective dreaming?

What does it mean if time is just a mental construct? If time is not ultimately real, then certainly there is no real past, nor future, nor even the present; nor can anything like a creation of the world have actually occurred.

Similarly, if space is an illusion, then the distinction between inside and outside—without which the world cannot be an objective reality—also becomes unreal. And then all the multitudinous limitations which have always appeared to pertain must also be illusory. For example, we feel separate from our fellow creatures. But if space is an illusion, this separateness must not really be true.

Chapter Four

Masters of Time and Space

Many wise, spiritual adepts throughout history have taught that the world is not as it seems. Patanjali, for instance, taught that it is possible to synchronize thought and action, putting them together so that no gap arises between them. When such a synchronization occurs, something truly amazing takes place—the limitations of space and time vanish. The individual to whom this occurs gains remarkable abilities and becomes a fully realized yogi—one with spirit. B. K. S. Iyengar, a devotee of Patanjali and founder of the Iyengar school of yoga, explains that with practice certain effects or powers known as *vibhutis* manifest for the yogi. These effects, documented by Patanjali in the *Yoga Sutras*, comprise knowledge of past and future, including past and future lives; the ability to "read" other people's minds—even precise details in thought; full awareness of the ways consciousness works; the ability to become invisible; and even the ability to levitate.[1]

Another recent master, who departed in 1950, exhibited many of these abilities. Born as Venkataraman in a village in South India, he later became honored in India, and by Westerners who knew him, as Sri Ramana Maharshi (see figure 4.1). It is certainly useful to look into Ramana's amazing realizations to learn what we can about how he achieved, through his mind-yogic practice, a means for time travel. When Ramana was twelve years old his father died, and he went to live with an uncle in the nearby city of Madurai. In 1896, at the age of sixteen, he left his uncle's home and journeyed to one of South India's holy sites, the sacred mountain called Arunachala. He had heard of this mountain briefly when one of his visiting elderly relatives had happened to mention it. Merely hearing the name "Arunachala" had a magical effect on the boy, generating an internal excitement which he himself could not understand.

A few months after hearing of Arunachala, and before he departed for the place, Ramana had an experience that changed his life irrevocably. Seated in a bedroom on the second floor of his uncle's house, he was suddenly overwhelmed with the fear of death and became fully convinced that death was imminent. This inexplicable feeling persisted even though he was completely healthy. Shaking with fear, he began to ponder the significance of his death. Since he was alone in the room, he decided to act out his own death and inquire into the meaning of it. He laid down with his arms stiffly at his sides, held his breath, and said to himself:

> Now death has come but what does it mean? What is it that is dying? The body dies and is carried off to the cremation ground and reduced to ashes. But with the death of the body, am I dead? Am I the body? This body is now silent and inert but I feel the full force of my personality and even the voice of the "I" within me, apart from the body. So I am the Spirit transcending the body. The body dies but the spirit that transcends it cannot be touched by death. That means that I am the deathless Spirit.[2]

How should we interpret this experience? One could argue that the youth was full of imagination and, through his ability to fantasize, produced the sensations of fear and imagined his death coming. However, from all later accounts, he was actually in a near-death state at that moment, even though he was physically healthy. Can the mind produce such a state? The awareness of his "impending death" took full possession of him, not merely as an idea but at a deeper level that opened up his spiritual self-awareness. He suddenly became spirit and knew himself as that, no longer identifying himself as merely the body form that had been called Venkataraman. His self-realization was instantaneous, complete, and irreversible. His ego was lost in a flood of pure self-awareness.[3] Soon after this experience, he left his uncle's home for Arunachala.

When he arrived at the mountain, he got rid of all of his possessions, including clothes, money, and food, and abandoned himself to his newly discovered awareness—that he was formless, spaceless, and timeless, and simultaneously everything at once.

Living in the natural surroundings of the mountain—dwelling in a cave on its slope—he became oblivious of the world, his immediate environment, and even his body. He was so absorbed in this quest that he didn't feel insects gnawing at his limbs, nor was he aware of the great lengths to which his hair and nails grew. For two or three years he remained in this state, living on whatever people gave him to eat.

Even though he gradually returned to normal appearance and bodily health, his mind remained fixed on his true nature as spirit. The Indian philosophers would call him self-realized—a fact obvious to all who came in contact with him. People immediately felt his spiritual radiance and would gather around him, even though he rarely spoke to them; just being in his presence was enough. Soon he was given the name Bhagavan Sri Ramana Maharshi. *Bhagavan* means Lord; *Sri* is a respectful term of address, something like *Sir* in the title of a knight; *Ramana* is a contraction of his name Venkataraman; and *Maharshi* means great seer (literally *maha-rishi*) in Sanskrit. He was also known as the sage of Arunachala.

Ramana explained to those who inquired that it is the mind that is vast, not the world. The knower is ever greater than the known, and the seer is greater than the seen. That which is known is contained within the knower, and that which is seen is in the seer; the vast expanse of the sky is in the mind, not outside, because the mind is everywhere and there is no outside to it.

From Maharshi's thoughts we learn that forms and the space that surrounds them appear only because of the ego-sense. As countless spiritual leaders before Ramana and many after affirm, the primary ignorance, out of which all external phenomena appear to arise, is the ego-sense. Going beyond this sense becomes a difficult task. You need to accept the role of primary ignorance and its limitation and see how the world of your own ignorant mind sets into motion a limited way of fear connected to your belief that you are confined by space and time. The sage described the steps to going beyond the ego-sense, which means going beyond space and time, and by doing so learn to time travel. One needs to direct one's thoughts in the following manner:

> Distracted as we are by various thoughts, if we would continually contemplate the Self, which is Itself God, this single thought would in due course replace all distraction and would itself ultimately vanish. The pure Consciousness that alone finally remains is God. This is Liberation. To be constantly centered on one's own all-perfect pure Self is the acme of Yoga, wisdom, and all other forms of spiritual practice. Even though the mind wanders restlessly, involved in external matters, and so is forgetful of its own Self, one should remain alert and remember, "The body is not I."[4]

IN THE SELF THERE IS NO SPACETIME

There are many stories illustrating Maharshi's remarkable abilities to transcend space and time. One of them concerns a couple from Peru who were visiting his ashram at Arunachala. One day they had the opportunity to speak at length to him. They told him their "hard luck" story, not leaving out the privations they had undergone to make the trip to India and visit him. Bhagavan listened patiently to their story with great concern, and then remarked: "You need not have taken all this trouble. You could well have thought of me from where you were, and so could have had all the consolation of a personal visit."

Listening from the place where most of us identify ourselves—the ego—the couple did not understand this remark; nor did it give them any consolation, even though they sat at Maharshi's feet in apparent devotion. Seeing that telling their story had somewhat calmed them, and not wanting to disturb their apparent "devotional" pleasure in being in his immediate vicinity, he left them.

Later, in the quiet of the evening, Sri Maharshi spoke further with the couple, and naturally their talk turned to their life in Peru. They began describing the landscape of Peru, in particular, the sea coast near their town. Just then Maharshi remarked, addressing the husband: "Is not the beach of your town paved with marble slabs, and are not coconut palms planted in between? Are there not marble benches in rows facing the sea there, and did you not often sit on the fifth of those with your wife?" They were astonished. How could Sri Bhagavan, who had never left Arunachala, know so intimately such minute details about their home and its environment? Sri Maharshi only smiled and remarked: "It does not matter how I can tell. Enough if you know that in the Self there is no Space-Time."[5]

Figure 4.1. *Sri Ramana Maharshi.*

Ramana Maharshi's presence graced the renowned sacred Arunachala hill during the first half of the twentieth century. He was known throughout India and to many in the rest of the world as the silent sage whose peaceful presence and powerful gaze changed the lives of those who came into his presence. In silence he radiated peace and contentment like a powerful beacon, effecting a change in anyone who came within his sphere. He encouraged people to look within and discover whether they are actually the body or the changeless, eternal self. His powerful example and influence led many people to experience this inner self as the same Self behind all awareness, above the transient mind, emotions, and body.

Here is the essential key that opens the lock to time travel: Find ways to suspend your bodily awareness. This can be accomplished in two paradoxically different ways. One way is to do as Maharshi did—contemplate your essential Self by holding one thought in your mind, "I am not the body." The second way involves practicing mindfully what Patanjali calls *asanas,* the physical poses and reposes used to bring the mind to concentration and absorption. As Iyengar explains, this often involves perfecting the asanas, going beyond the trials of learning the poses to discover both the power and the limitations of the finite body. That is, by becoming fully conscious of the body, the practitioner realizes that he or she is other than the body and merges with the soul.

What's Time Got to Do with It?

Certainly Ramana was able to reach beyond spacetime limitation. The question remains: How was he able to do this? Was he merely escaping the trap laid by the metaphors that convince each of us that we cannot go beyond time and space? Here we'll look into how physics has brought the meaning of time into question.

In physics, time has different meanings depending on how the researcher carries out what's called a "program"—a procedure to investigate, using a mathematical theory, and measure some physical property of a phenomenon more fully. All programs ultimately involve three aspects or measurements. These are: Measurements of space—where does the phenomenon occur in relation to myself? Measurements of mass or energy—how much matter or energy is present? And measurements of duration—

when does the phenomenon occur, and how much time can I ascribe to it?

Space and matter are separately measurable and remeasurable in and of themselves; but time, surprisingly, is not. In fact, although this sounds strange, time has no absolute meaning in these programs, in that there is no way to determine in any absolute sense the time when a measurement occurs. In other words, when it comes to measuring time, even though we all know what a watch does, and even what an atomic clock does, we don't really know what we are measuring because we can't compare any one measurement with anything else.

Measurement always involves a comparison of the thing measured with a thing known. We know B and we measure A. Then we compare: Is B larger, smaller, or the same as A? Usually we have B in hand, as for example when we measure the length of a table with a tape measure. Here the tape measure itself provides the needed B. But I can't, for example, take the second that passed between 11:59:59 P.M. and midnight Pacific standard time on December 12, 1931 and compare it with the second that passed between 5:05:04 and 5:05:05 A.M. on January 12, 2001, central standard time. In fact, I have no second in hand regardless of when it occurred. I can't even compare the second that just passed with the one I am experiencing now.

Of course we use standard clocks continually to tell what time it is, such as those that emit time signals from places like the National Bureau of Standards in Boulder, Colorado. We receive a beep followed by another beep and infer that the interval between beeps is a single second, for example. But we never get our hands on that interval. I may have a concept about these two different seconds, but since those times are not "present" here and now for me to perceive, they remain immeasurable.

All past times remain so. All future times also are immeasurable, since they, too, are not here at hand, as it were. We can anticipate the future time only in our imaginations, and the past times are only in our memories—if we have them, and if not, past events too are only experienced in our imagination.

So that leaves just the time you are experiencing now as a possibly measurable quantity. But the instant you claim to measure it, by, say, looking at the watch on your wrist, that time mysteriously vanishes into the great depths of the immeasurable past. Hence, has any time actually ever been measured? Has any time really ever "passed"?

Take a moment now and watch your clock's sweep-second hand, or if you have a digital watch, take note of the flickering number changing. Try to be still and just observe; you should perceive, with a little practice, that nothing has actually moved or changed.

Thus there really is no present moment—or no past moment or future moment—ever present at all. It appears that time is moving, but if you actually look close enough, what you'll see is one thing vanishing and another appearing. Your mind puts these vanishing and appearing acts together and connects them, and in so doing provides you with your first and primary illusion of continuity, meaning the semblance of past, present, and future. But as with any good illusion, don't fall for it; it is your trickster-mind fooling you into believing in the persistence of time. However, there is more to this trick. If time is really an illusion, then what about space? Is space also illusory?

What's Space Got to Do with It?

We generally infer the existence of space because of movement—we observe something going from here to there and infer that it is moving through space. Without the movement of something or other, we wouldn't think in terms of space at all. Consider, for instance, how you measure space or decide that something occupies space. You "try it on" or "try it out"—for instance, when you determine how much space a shoe should provide to enclose your foot comfortably, how many inches it is on the tape measure from one side of your living room to the other, or how far it is to Aunt Minnie's house two blocks away. When we consider distance, we often think in terms of time as well; for example, Aunt Minnie's house is just ten minutes by foot.

When it comes to astronomical distances, movement is crucial for deciding how "big" space is or, for that matter, how far one thing is from another. You may peer at the stars and say to yourself, Wow, that star must be far away because I certainly can't reach it with my arms! As "arms" to reach the stars with, astronomers use the light they observe coming from those stars. They also, at times, use devices based on laws of mathematics—specifically trigonometry—to determine how far away something is. Using such extensions of their own arms and eyes, and their logical minds, they infer great stellar distances, since they can't drive to a star's address in the family car. Nevertheless, without movement of one thing or another—in the case of stars, the movement of light—we would have no idea how "big" space is.

Movement is crucial to our grasping of the concept of space, but when it comes to how we experience whether something moves or not, we must also use our minds. This is where psychology comes in.

PSYCHOLOGICAL SPACETIME

Psychology remains a remarkable field when you bring it together with psychophysics and physiology. In an interesting article published in 1992, philosopher of the mind Daniel Dennett and his associate, Marcel Kinsbourne, describe a number of paradoxes associated with the timing of conscious experience.[6] One of these is an apparent-motion phenomenon now known as the *color phi* effect, which was thoroughly researched and tested in the 1970s by physiologists P. Kolers and M. von Grünau.[7] The phi effect is quite amazing when you first hear about it and begin to think about its consequences.

Here is how it works: If you observe two small spots of light flashed in rapid succession on a facing wall and separated from each other by as much as four degrees of visual angle (the angle formed by drawing straight lines from both spots to just one of your eyes), it will look as though a single spot is moving from one location to the other. Although the phenomenon occurs with more than two spots flashed one after the other, for the sake of our discussion we'll keep the example to just two spots.

To grasp this better, imagine yourself looking at a blank wall. First one spot flashes on the wall and then extinguishes. Shortly thereafter the second spot flashes, perhaps a little to the right of the first spot. If the second spot follows the first rapidly enough, you will see the phenomenon of a single spot moving from the left to the right continuously, as if someone were shining a flashlight on the wall in a short horizontal line. To really carry out this experiment you would need to be able to flash the spots on the wall very quickly with one spot following the other in less than one fifth of a second. The persistence of this illusion is why we see motion in movies and on television; otherwise, we would see only flickering images.

After Kolers and Grünau had carried out this and similar phi experiments, the philosopher Nelson Goodman asked them to test what would happen if the two spots were of different colors (any two colors would do), say, red and green. That is, the first spot flashed on the wall would be red, and the second one, flashed an instant later and to the right of the first spot, would be green. How would anyone (with normal color vision) experience these spots?

We might guess that the observer would see intermediate spots along the trajectory gradually change color from red to orange and so on through the color spectrum until it became green. Another guess is that the presence of color would make the illusion of motion vanish, leaving behind only the images of two differently colored flashing spots.

What is actually perceived is quite amazing. Kolers and Grünau did the experiment and were astonished to find that the apparently traveling spot remained red until it got about half way along its illusory trajectory, and then it suddenly flipped to green before continuing to its final location on the wall. Think about this. No light has actually been flashed along the trajectory between the two spots. Not only is the movement from left to right on the wall an illusion, but the red switches to green *before* the green spot ever appears on the wall! How can this be?

Before the green spot appears, the person presumably has no idea what color it will be. It could be yellow, or even purple—a color not between red and green on the spectrum—for that matter. So he or she can't guess. The experimenters tested this using several observers and different colors. The subjects of the experiments did not know what the final spot's color would be. Yet they all seem to see the same change on the very first trial. The subjects reported that the color switched halfway along the illusory trajectory and always to the color of the second spot. You might suspect that the observers somehow construct the illusory path in their minds after the green spot appears. But why would their minds perform this rather elaborate trick? Does this mean that we never really see what's "out there" when it is happening, but only reconstruct the scene in our mind and then project the reconstruction back into time? If so, why bother to make up a trajectory in the imagination?

Does the brain act like a time machine, projecting the experience backward in time? Alternatively, does it experience some kind of precognition? Clearly, the illusion of the switch of color at the illusory midpoint cannot occur until *after* the second spot registers in some way in the brain. So the mind must look forward in time, see the green color, and then back up in time to make the switch.

The most plausible explanation has the observer projecting the experience back in time from the perspective of having already seen it. However, it would then seem to be too late and hardly necessary to interpose the illusory color-switching-while-moving scene between the conscious experience of spot 1 and the conscious experience of spot 2. How and why does the brain accomplish this sleight-of-hand? The logical explanation would be that all of this appearance-of-a-moving-spot-with-color-change was taking place after the final green spot flashed and not before.

But, considering the short time between the flashing and the describing, this hardly seems plausible. And what purpose would such a process have in the evolution of things?

Backwards in Time through Our Brains

There are other backward-through-time experiments in physiology. Ben Libet began researching how our brains appear to process data backward in time in the late 1950s. Working with brain surgeon Bertram Feinstein at Mount Zion Hospital in San Francisco, Libet, a neurophysiologist, began by studying the brains of patients in the operating room as they were undergoing surgery. So that Libet could continue to monitor the patients postoperatively, Feinstein implanted electrodes in their brains and left them in for some period of days. By the time of Feinstein's untimely death in 1978, Libet and his associates had completed the work for his now-famous 1979 paper on subjective referral.[8]

The astonishing fact Libet discovered can be said quite simply: We are mostly unconscious. That is, we make decisions and respond to sensations from the outside world unconsciously. We only become conscious of the actual stimulus much, much later, after the slings and arrows of our fate have already passed us by, or struck us. But there is an interesting twist: When we think about what just happened, we refer back in time from the later moment of conscious awareness to the earlier moment of sensation; we also refer out in space to the location of the stimulus, even though our actual perception occurs in our bodies. As we shall see, this back-in-time, or temporal, referral and out-in-space, or spatial, referral provide a profound basis for using the mind to time travel.

Spatial referral has been known about for some time. For example, if the brain of a subject is stimulated in a particular area of the cortex, the person will have certain sensations in the body, such as feeling a pin prick on the forearm or hearing a sound. We all experience objects "out there" in space. We see cars on the street. We hear horns honking. We smell baking bread. We never doubt that these sensations arise in our bodies, yet we refer them all to the objects "out there." Yet the mechanisms by which visual images are reconstructed, and to which we are actually responding, are located within our brains, neural networks, and retinas. In a similar manner, we reconstruct from the vibrations of our ear drums the approximate location of a sound's source in space. That is why we look up when an airplane passes overhead. This is called subjective referral in space.

Libet's experiments had to do with temporal referral. He carefully applied physical stimuli to the bodies and, through Feinstein's attached electrodes, electrical stimuli to the brains of several subjects. He expected to find some small delay between a person's conscious awareness of some stimulation event and the time when a stimulus is applied, but it was assuredly a surprise to find that the delay time was typically as much as a quarter of a second, even half a second. A lot happens in a half-second. A 90-mph baseball leaves a major league pitcher's hand and crosses home plate 60 feet away in less than a half-second (in 400 milliseconds, which is 400 thousandths of a second), and yet the batter manages to hit the ball accurately, getting a base-hit 20 to 33 percent of the time. An animal darts out into the street ahead of your car, and you manage to hit the brakes in less than a half-second from the time the light from the headlights reflects from the animal to your eyes. There are many other equally ordinary examples in which, Libet would argue, the person is totally unconscious at the time of response. That is, the reaction is faster than the perception.

Libet refers to his theory of consciousness as "subjective antedating" or subjective referral in time. His data show that a person, although able to react to stimuli within a hundred-thousandths (100 milliseconds) of a second, is not actually aware of what he or she is reacting to for several hundred milliseconds, up to a full half-second. Yet when asked just when he or she became aware of a stimulus produced by a certain event, the person responds as if he or she were aware at the time of the event itself.

A good example would be a 100-meter runner at a track meet. He leaves the starting block around 100 milliseconds after the starter has fired his gun. But it is not until some 250-400 milliseconds later that he actually perceives the gun shot. By this time he is well on the way towards the finish line, perhaps five meters down the track. Yet if we ask him later about his experience, he will say he was conscious of the shot at the time he pushed off from the block. Amazing as it may sound to us, Libet claims that it is not possible for the runner to be conscious of the shot even though he responds as if he were.

Although there are complex explanations, this temporal referral paradox and the two-spot color phi paradox remain mysteries to this day[9]—if we are committed to a linear concept of time as the basis for understanding reality. As I see it, however, these are not really paradoxes but indications of how time and consciousness work together suggesting some new metaphors. We will look into this more deeply in the upcoming chapters.

Why Does Time Go Forward?

Connected to these paradoxes is another question, one that causes great ripples of concern across the brows of philosophers of time. Again to use a metaphor, why does time go forward? It turns out that the forward motion of time is an illusion rooted deeply in the Western psyche, founded in one single concept—that time is linear.

If time is linear, it must have a direction, an arrow, so to speak, and we refer to this arrow to mark how one experience differs from another in an essential way. But even though we assign a direction to this time-arrow, is it inevitable that all experiences should follow in that direction? Does the arrow of time act like a highway cop, directing all who attempt to move counter to the flow of traffic to move with it?

There are two basic notions about the movement of time. One of them—a view mostly found among ancient peoples—acknowledges that cycles of time faithfully repeat themselves (indeed, a cycle means a repetition) and hence are replays of an original succession of events. The other, which most likely came about through the invention of mechanical devices such as the steam engine, recognizes that even though cycles appear to repeat, they are not identical; something about them fails to exactly replicate what preceded.

LAWS OF NATURE AND LINE-TIME

People in the nineteenth century were enamored of machines. You might suspect that this love affair began in the seventeenth and eighteenth centuries with Sir Isaac Newton's discoveries of the laws of mechanics, a remarkable set of mathematical equations describing how imagined concepts such as mass, velocity, force, acceleration, work, and energy can be associated with real human experiences.

Even though Newton is credited with inventing the so-called clockwork universe, it is doubtful that Newton had much experience with clocks, for in his day, they were cumbersome devices, and few people ever even saw one. A clock would have been about as common to an eighteenth-century person as a particle linear accelerator is today.

Nevertheless, mechanical principles made their way into common life through the Industrial Revolution, and by the nineteenth century people began to notice that steam engines—the most common industrial engine before the discovery of electricity—followed a basic tenet that came to be known as the first law of thermodynamics. This law says that something called energy is capable of undergoing complete transformation in the operation of a machine. People became aware that something called "energy" must really exist, and when a steam engine operates, that energy can be changed into work or heat. But ultimately, energy can't be created out of nothing, nor can it be destroyed.

Energy became a means by which the industrial outputs and inputs could be measured and accounted for. The term "work" when applied to human labor was no mere analogy but a true representation of mechanical law. It was directly related to a change in the energy in any machine, including the human body.

In the case of steam engines—and the devices that used them to mechanical advantage, such as pushing or pulling a locomotive along a track or raising a weight—the energy existed in the chamber holding the steam. The work done by this energy appeared whenever a piston connected to the steam chamber was moved outward by the steam, in turn causing the wheels of a locomotive (or other such device) to turn, thus moving it along a track.

People noticed that the steam engine would eventually run out of steam, unless someone added fuel to the fire so the chamber would maintain roughly the same energy as it had before it expanded the piston. Thus it became clear that heat is a form of energy. People also noticed that the turn of the wheel eventually causes the piston to move back in again, compressing the hot gas. This heats the gas back up, seeming to restore its energy. So why didn't the gas just push out again with the same amount of oomph?

Why was it necessary to keep the fires hot to keep the steam energy up? Why wasn't the energy in the chamber the same after a completed cycle of the piston? People soon realized that when the locomotive moved along the track, with its wheels rubbing against the steel rails, it was resisted by the air it pushed aside. Since pushing air and rubbing against steel rails also took energy, people figured out that fuel is needed to supply energy for these phenomena, too. This wasted energy was eventually given the name "friction." Some amount of the energy goes into friction.

Why can't we get back that friction energy from the air and from the rails? If the energy moves from the hot chamber eventually to reach the cold rails and the relatively cooler air, why can't we retrieve it from the rails and the air?

It can't be done. People recognized that there was some other principle of nature involved, one that surpassed Newton's vision. This law is the second law of thermodynamics, and it says, in simple language, that you cannot take heat energy from a cold body and heat up a hot body unless you do some work.

Hot bodies cool off, while cold bodies in contact with them get warmer. No one ever saw the reverse—cold bodies getting colder while their hotter neighbors who are touching them get warmer. Recognition of the one-way flow of energy ultimately led to the direction of time's arrow, the concept of the unidirectional flow of time. If a cold body and a hot one were in contact, and if nothing was added and no work was done to change things, then the observation that the hot body got cooler and the cold body warmer told people that it had gotten later—that time had passed. Today this observation is known as the thermodynamic arrow of time. It tells us that time does not go in cycles but continues along a track, just like the locomotive.

Hence, even though an engine has a repeating cycle, nothing ever repeats exactly. Something is different at the end of each in-and-out motion of the steam piston. The fuel heating the "boiler," as it came to be called, is diminished after a single cycle, and it continues to diminish.

Put the thermodynamic arrow of time together with the energy required for manual labor and you come up with the time-clock punch—the means to measure the working person's wage. Linear time thus became the ultimate frame upon which Western culture determines its technological progress, its labor laws, and its riches, or lack thereof.

Sacred Laws and Hoop-time

Do we experience time directly? We have many concepts and images of time, but these are not time itself. Let me refer to your common experiences, right now. Do you experience time? If you take a moment and look deeply within, you will see that the answer is "no." You may disagree. Fair enough. Then tell me about your experience—or I can tell you about mine.

As I sit here typing out these words I can look about the room, and I believe I can experience a sense of the space. By moving my eyes up, to the right and the left, I can experience the walls of the room and the ceiling. That is, I see these things and I believe they are there. I can experience the mass of the objects I touch. I can feel the force of the seat against my backside and the force of the floor upon my feet. I can also feel the force of the back of my chair pressing against my back. In this manner I believe I am experiencing mass and locating my body in space.

But what about time? All I can really experience is the instant when each event I am conscious of occurs. In that instant, I can also experience the sense of memory and the sense of anticipation—the feeling that I have done this before and that I will do it again. But in none of these do I find time. I certainly can infer time by watching a clock or even watching my leg as it taps on the floor in a rhythmic pattern. Indeed, music is one way for me to infer time—through my sense and appreciation of rhythm. But time all by itself? It simply isn't sensible—I can't feel it or see it.

If I can't sense time, then why do I believe it moves forward?

Let's take a moment to revisit the aboriginal concept of time. Aboriginal time appears as a rhythm or cycle—I think of it as a sacred hoop that, for the Western mind, can be pictured as rolling along and touching line-time at every instant. Hence the direction of time becomes immaterial; it simply doesn't matter what is past, present, and future. The important thing is the presence of the hoop touching life, as indicated by the line of time it touches at every moment.

Mircea Eliade writes about the aboriginal view of the coming into being of the world and mankind in the Dreamtime.[10] The physical landscape was changed and humans became what they are today as a result of a series of deeds by supernatural beings. Yet today nowhere in Australia do these dreamtime personages impress us with their grandeur. Rather, the majority of the central

Australian creation myths tell only of their long and monotonous wanderings. When these supernatural beings, born of the earth, had accomplished their labors and completed their wanderings, overpowering weariness fell upon them. The work that they had performed had taxed their strength to the utmost, and they sank back into their original slumbering state. Their bodies either vanished into the ground—often at the site where they had first emerged—or turned into rocks, trees, or sacred objects.

The Dreamtime apparently came to an end when these supernatural beings became the earth. But the mythical past was not lost forever; to the contrary, it is still periodically recovered through the tribal rituals. In this way the rolling hoop of sacred time enters consciousness. Initiates today learn how to relive Dreamtime through ceremonies. Eventually the initiate becomes completely immersed in the sacred history of the tribe, absorbing the origin and meaning of everything from rocks, plants, and animals to customs, symbols, and rules. As the initiate assimilates what is preserved in the myths and rituals, the world, life, and human existence become meaningful and sacred, for he or she understands that everything has been created or perfected by supernatural beings.

At a certain moment in their lives, initiates discover that before their birth they were spirits and that after their deaths they are to be reintegrated into that prenatal spiritual condition. They learn that the human cycle is part of a larger cosmic cycle, that creation was a "spiritual" act that took place in the Dreamtime, and that although the cosmos is now "real" or "material," it nonetheless must be periodically renewed by reiteration of the creative acts that occurred in the beginning through ritual. This renovation of the world is a spiritual deed, reinforcing communication with the eternal ones of the Dreamtime.[11]

In the context of the ritual, a sense arises that the time of the Dreamtime need not be in the past, nor do the events of the Dreamtime need to fit within the scheme of passing time. Instead, the Dreamtime and ordinary time fit together as two contrasting yet interrelating temporal images.

If we were to describe the time when the Dreamtime was in full bloom, we would think of it in the past—the proverbial long time ago.

Figure 4.2. *Line-time and Dreamtime.*

In figure 4.2, the letter *D* at the beginning of the line represents the origin of a Dreamtime event. The letter *N* represents "now," the line stretching between *D* and *N* represents the past, while the line to the right of *N* represents the future. Using this line-time model, I would say that Dreamtime was a long time ago.

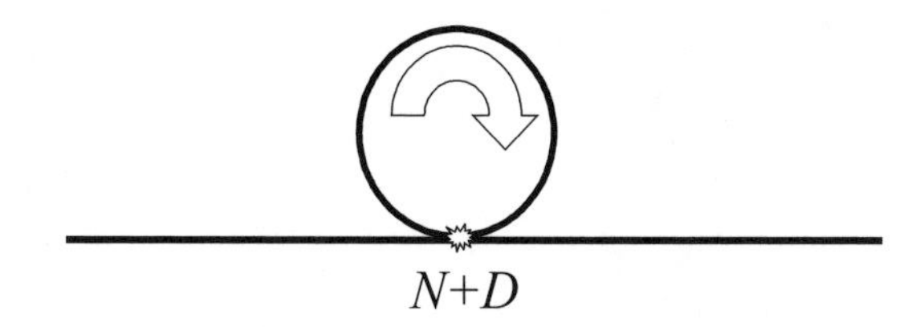

Figure 4.3. *Hoop-time and Dreamtime.*

Now consider Dreamtime as a sacred hoop of time (figure 4.3).

If we think of it as rolling along line-time, we see that Dreamtime is ever present: It was in the past, it is in the present, and it will be in the future. This is the nature of mythical time. All events in the "past" are equally present. That they are not to be taken as events of chronological time means that they are not to be put on a time line, as we might do with the events of our lives. This doesn't mean that they are not real or didn't happen, or for that matter that they are not happening now.

In this way, sacred time brings line-time alive. It provides the spark that comes from the hoop touching the line. Because our memories have a lot to do with these sparks, the direction of the rolling hoop gives a direction to temporal experience.

5 CHAPTER FIVE

The PHYSICS *of* "ORDINARY" TIME TRAVEL

Time is a fire that consumes me,
but I am the fire.
—Jorge Borgès

In this chapter, we shall take a tour through ordinary space, time, energy, and time travel as seen by modern physics. I use the word *ordinary* to refer to the usual or accepted way that modern physicists think about time travel. In chapter 8 we will look at what could be called extraordinary time travel, which builds on what we know from quantum physics but incorporates wisdom of the mystics. It may seem strange to call any kind of time travel ordinary; however, you have to admit it's a familiar concept, one that has been in the Western mindset as a possibility for more than a hundred years. Certainly science fiction is full of examples. Perhaps one of the earliest, if not the first, story to deal with time travel was Herbert George Well's first novel, *The Time Machine,* published in 1895. Wells's vision catapulted him into the public eye. Although he preceded Albert Einstein and Hermann Minkowski (see chapter 3) by more than ten years, he somehow foresaw the coming "mystical physics" age that would arise out of Einstein's strange story of space and time.

Chapter Five

The First Time Machine

Wells was apparently fascinated with the idea of dimension. What a physicist means by dimension may not be quite what you have in mind. Dimension actually refers to freedom—the freedom to move. For example, you can move along one dimension—right and left—you are free to do so. You can move along a second dimension—forward and backward—with equal freedom. With the help of modern conveyances, you can move along a third dimension—up and down—with nearly the same amount of freedom, though in this case your weight and the force of gravity hinder you somewhat. Physicists propose that time comprises a fourth dimension, although it seems we can't move in it as freely as in the others, unless perhaps we have the proper conveyance—a time machine of some sort.

Wells introduced another idea here well before its time—the concept of "temporal thickness." Having a certain thickness in time means that our human experience can't really be narrowed down to infinitesimal instances—one following the next, in order. Rather, each instance must be "fuzzy," or spread out, occupying an extent of time. With these concepts in mind, we will look into Wells's book.

In the opening chapter, Wells's lead character, named simply "the time traveler," explains to his friends and colleagues that he has come to a remarkable conclusion: School geometry is founded on a misconception, namely, that geometric figures exist. They cannot exist because they are constructed out of objects that do not exist. It all comes down to the simple point. A geometric point has no height, width, or length, hence cannot exist. Therefore, a line—which is nothing but a series of, in fact, nonexistent points—cannot exist either. Similarly, a plane or a cube cannot really exist because they are made up of lines. Therefore, no geometrical object can be real, simply because it is constructed of smaller objects that are themselves not real.

Space and Time Have Thickness

Of course his friends protest and argue that objects like cubes certainly do exist. The time traveler, after a silent pause, asks his friends, notably Philby, who plays his foil, "Can a cube exist instantaneously?" Philby is confused by the question, so the time traveler explains by asking another, "Can a cube that does not last at all, have a real existence?" As Philby ponders, the time traveler explains, in essence, that for a *real* point to exist "out there" in the physical world, it must have some "thickness."

What is thickness? It's the capacity of an object to register presence in the next dimension up. Real objects do this, but geometric objects do not. For instance, a geometric line conceptually has one dimension—length—but no width, so it has no thickness into the second dimension. An actual line drawn on the page, however, does have some breadth—it has some thickness—or we couldn't see it. The same principle applies for a plane and a cube. It must follow, therefore, that a real cube can only be so if it has some presence in the fourth dimension, that is, a "temporal thickness"—a duration.

By this logic, all real things in our three-dimensional world, regardless of how we think about them abstractly, must exist in three spatial dimensions plus have a thickness spreading from them into the fourth dimension, time. However, we humans fail to see the temporal thickness of three-dimensional objects because, as the time traveler explains, "Consciousness moves intermittently in one direction along [time] from the beginning to the end of our lives." That is, we can move in only one direction in the fourth dimension—along time into the future—whereas we can move in two directions in the first three. This gives us the ability to scan the first three dimensions and to go back and repeat any measurement we would make of them, but not to do so in the fourth dimension.

The friends in Wells's story begin to discuss whether it is possible to travel willfully—that is—freely, in time as easily as we travel willfully in space. Philby of course says no; willful traveling in space is easy, one can go right or left, forward or backward, and up or down with ease. But time travel cannot be so easy.

The time traveler points out a missed step: Willful traveling in the first two spatial dimensions is easy enough, but what about going up and down? To ascend to great heights in space requires the use of a machine, for instance, a hot air balloon.[1] On the other hand, descending is easy enough—the tendency to fall governs all of us. Thus the third dimension, too, is limited when it comes to free movement. We can move in one vertical direction more freely than in the opposite. But with the help of devices of one kind or another, we can move freely and willfully in that direction, too.

Even though there appears to be some kind of field opposing our free movement in the third dimension, we are able willfully to overcome it through technology. We understand that field to be gravity, and we can construct antigravity machines, such as hot air balloons, to allow us access to previously inaccessible regions along this dimension. The time traveler then tells his friends that he has built a proper device enabling one to move just as willfully in time—in the other direction, backward through it.

Wells thus suggests that there may be a field acting in the fourth dimension, tending to keep us moving toward the future, just as gravity acts in three-dimensional space, tending to keep us moving downward.

Fuzzy Reality in a Frozen Block

Wells continues this field theme as he has the friends consider space and time as "frozen" dimensions. A three-dimensional block of ice, we say, exists in three-dimensional space. Comparably, an entire human life, which exists over time, can be conceptualized as a four-dimensional tube set in a four-dimensional block of spacetime. The time traveler explains that each of us exists as a solid entity in four dimensions. Every person has fuzzy thicknesses extending in all dimensions and consciousness precipitating out, instant by instant, like a dew drop from a fog, with each drop a fuzzy cross section of the person's whole life. Although we exist as four-dimensional beings frozen in spacetime, we only experience the precipitation process, moment by moment. Consciousness moves along only one of those dimensions—the time dimension—and only "sees" a cross section of the whole, which it takes to be the whole being at a certain time. If you look through a photo album of photographs of your grandfather, you will probably find photos of him at various stages of his life. A baby picture is a two-dimensional representation of him near one end of his life-tube, while a portrait of him as a young man is another cross section of the tube, and so on. We can think of these as frozen moments of time—a four-dimensional event recorded in a two-dimensional medium so we can handle them. When you go to your grandfather's house and meet him in person, you are in one sense meeting a momentary, three-dimensional presentation of the four-dimensional person, in terms of his entire lifetime of experiences. That is, a living person would be the whole tube extended in four dimensions, so that an actual baby appears as a three-dimensional cross section through the tube, a young man as another cross section, and an old man as another, taken at the end of the temporal dimension for that particular individual.

This notion of "block spacetime" is pervasive in modern physics theory arising from Minkowski's vision discussed in chapter 3. "Ordinary" time travel frees one from the "time tube" instantaneous cross section in four-dimensional spacetime and allows one to move along the temporal dimension as freely as one moves from one room of a house to another. To actually accomplish this, however, means finding some technology that can counter the natural tendency to move along the tube in a single direction—into the future—instant by instant as we all seem compelled to do.

Perhaps unwittingly, Wells also foresaw the connection of time to gravity. He points out that our difficulty in moving freely in both directions in the third, vertical dimension has to do with gravity being a field of force. In a force field it is easier to move in only one direction along it, namely, from a "thinner" toward a "thicker" region. When it comes to gravity, "thicker" means a region where gravity is stronger and so has a stronger influence; "thinner" of course describes the region where gravity is weaker and has less pull. The field tends to move us from a weaker field of gravity to a stronger one; that's why we "fall" in the earth's gravitational field.

Time also responds to this thin-thick quality of gravity. Clocks tick more slowly in a thicker gravitational field than in a thinner one. This is why clocks that orbit the earth in satellites 22,237 miles above us—the geostationary satellites, which rotate in synchronization with the planet's surface—actually tick faster than clocks directly below them on the earth.[2]

Gravity's Stretch Marks

After Einstein came up with his general theory of relativity, certain concepts that had made sense for many years needed to change. Even definitions of common terms had to change, for now we saw the world differently than before. Consider, for instance, the concept of gravity. We usually take for granted that gravity "holds" us to the earth. As children, we may have wondered about the kids living in Australia: How did they keep from falling off the planet? Since they clearly were upside down compared to us, shouldn't they all be tumbling into space? Then we learned that they don't fall off because, even though the earth is round, gravity is at work, so "up" means the same thing for them as it does for us.

In the pre-Einsteinian and post-Newtonian way of thinking, the earth is round, although not perfectly so, since it bulges at the equator because of the earth's rotation; gravity is a field of force, acting on all hunks of mass equally in all directions in an attractive manner, so it pulls things together that were apart. Whereas the simplistic notion is that the earth holds you down by gravity, in fact you and the earth attract each other with the same force—it pulls on you, and you, believe it or not, pull on it. Just the fact that your body has mass is all that is needed to create the attraction.

With Einstein's field equations, which comprise his general theory of relativity, our picture of gravity has changed. We used to think of gravity as a field of force—like a magnetic field that causes pieces of metal to move—and gravity is that field that pulls us to the earth. Since Einstein's discovery, which showed how matter and energy "stress" space and time, gravity is now seen as a distortion of spacetime. Consider the following visual metaphor: Imagine all of three-dimensional space squished into a great two-dimensional sheet of plastic held taut by stretching rods. Then envision putting a bowling ball on the sheet, and notice how it distorts the plastic most strongly where the ball rests on the sheet. Now imagine the ball rolling along the sheet and you'll note that the plastic begins to ripple. Waves even arise and travel along the sheet if it is taut and flexible enough.

Gravitational waves are like the waves in the plastic—distortions of space, really, induced by moving matter. Now imagine that the indentations on the sheet, including the big one made by the bowling ball, persist in a ghostly manner even after you remove the ball. As far as Einstein's equations are concerned, the same thing happens whether or not the ball is present. Wherever a deep indentation exists—making a "gravity-well"—mass is said to exist. Hence, mass can be replaced by gravity-wells, or distortions of space. Based on these field equations, we no longer envision the universe as made up of matter and space; we now regard it as made of "crinkled" spacetime.[3]

Gravity and other forces may indeed crinkle spacetime, yet this isn't the entire picture. Here, common sense and use of pre-Einsteinian terms can be confusing. A distortion of space *is* the same thing as a gravitational force, even though we might on occasion say mass distorts spacetime, as if mass causes the distortion. Following this line of thought, since mass makes a gravitational field, and gravitational fields distort spacetime, then mass distorts spacetime. We can interchange mass, gravitational field, and spacetime distortion in any order, so that for example we could say: Since spacetime distortion makes mass, and mass generates a gravitational field, then spacetime distortion makes a gravitational field.

So, and this is crucial, we don't need to envision mass or a gravity field separately anymore—the distortion of space is the same thing as either one. When the question arises, which of the three is the cause and which is the effect, we answer by saying none of the above. We use these concepts interchangeably, sometimes saying mass distorts spacetime, or perhaps poetically, mass is nothing more than gravitational stretch marks in spacetime, and so on.

Since space and time are now considered nonseparate, mass can be seen as a distortion, not only of space, but also of time. Such a distortion is much harder to illustrate, since we can't really see time. Physicists get around this problem by using one of the spatial dimensions to represent time—as in the example of block spacetime. Imagine, then, that one of the dimensions in the plastic sheet is a time line, but this time is drawn radially outward from some fixed point so that, as we move out along a radius from this point, time increases. Movement along a circle around the point on the plastic we imagine to be simultaneous movement in space. The radius and circle comprise a circular spacetime coordinate system (see figure 5.1).

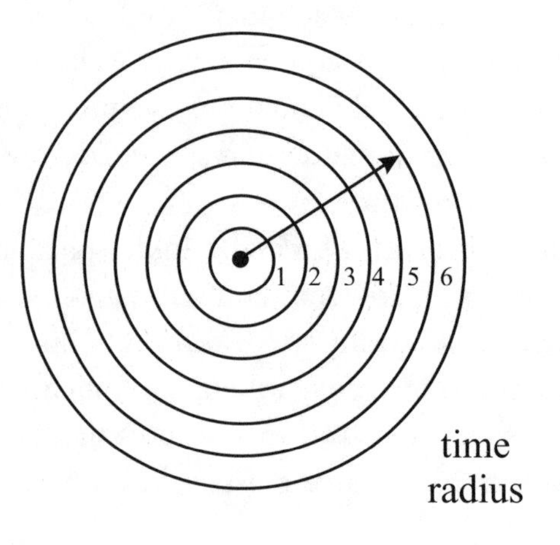

Figure 5.1. *Circular spacetime coordinates.*

Suppose We mark off equal durations of time on the plastic sheet at, say, one-second intervals. They would appear as a series of concentric circles extending outward from the central point. Placing the bowling ball on that point, we find the plastic around the ball stretching. As we move inward circle by circle toward the ball along a radial line, we see that each circle has increased its diameter proportionately a little bit more than the preceding circle, with the maximum distortion showing up where the ball comes in contact with the central point. That is, we see each second stretched out longer and longer as we move inward toward the ball's contact point with the plastic sheet (see figure 5.2).

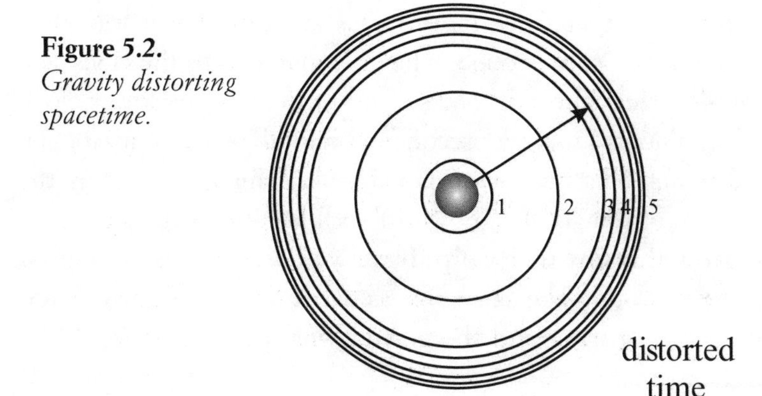

Figure 5.2.
Gravity distorting spacetime.

Time is stretched—dilated, or slowed down—in a gravitational field. The stronger the gravitational field, the greater the slowing of time. When we apply this principle to the earth's gravity, we can then begin to think of gravity as curvature of time! In other words, gravity can be related to a time warp—a distortion in the movement of time that occurs every instant when you move up or down—for example, when you move from a higher to a lower floor in a highrise building.

Can we measure the difference in time between two vertical locations on the earth? Yes. The effect of gravity on clocks was first measured by physicists R. V. Pound and G. A. Rebka at

Harvard University in 1959.[4] This was also the first time a time warp was measured. Pound and Rebka set up two timekeepers, one in the basement of a building on the Harvard campus and the other seventy-four feet above, in the building's penthouse. According to the general theory of relativity, the two clocks would not keep the same time because of the spacetime curvature produced by the gravitational field. Next the two physicists figured out how to send a timing message from the basement to the penthouse. This message would carry a very small time interval in it. This time interval matched the clock rate in the basement, but, because of the time curvature due to the change in the gravitational field, would not match the clock in the penthouse.

How large was the time warp measured by Pound and Rebka? The effect was quite tiny; indeed, it is amazing that it was measured at all. But sure enough, when compared with the clock just seventy-four feet above it, one second of time on the lower clock lasted 5 millionths of a nanosecond (a billionth of a second) longer.

Although this time warp is extremely small, it gives us the experience of gravity. The powerful pull holding us to the planet is made by this tiny time warp. If we were living on a more massive planet around the same size as earth, the time warp would become even greater and the gravitational pull on us would be even stronger.

Pound and Rebka's experiment confirmed the slowing down of time as predicted by Einstein's geometrical spacetime curvature theory. Moving upwards from the earth's surface in a highrise building to the twentieth floor, for example, we move into less spacetime curvature and therefore more rapidly ticking clocks. All of our physical processes would speed up. We walk around this planet oblivious of the fact that our heads are running slightly faster than our feet. It seems that we *live* on a time machine, as perhaps Wells was foreshadowing in his story.

Can we slow time indefinitely, freezing it altogether, by moving into an infinitely thickening gravitational field? It appears that the answer is yes again. But in fact, this slowing down seems to reach a limit; space and time get so thick that they produce a "fold" in the spacetime fabric called a black hole. Black holes have very strong gravitational fields and hence can slow time considerably, especially as we approach what is known as the black hole horizon—a place where time as we know it appears to stand still and even light appears to come to a screeching halt.

Falling into a Black Hole

A black hole is thought to form when a massive dying star, after burning up all of its nuclear fuel, collapses under its own weight into a single point called its essential *singularity*. In the early to midsixties, black holes were theoretical curiosities predicted in the obscure equations of the general theory of relativity. In today's cosmos, astronomers see black holes at the centers of every galaxy, so they have passed from curiosity to fact. Astonishing as this prediction remains, even more unusual occurrences are predicted to occur within the collapsed star's boundaries.

The first to realize the bizarre consequences of the implosion of a heavy star was the German physicist Karl Schwarzschild, who actually solved Einstein's field equations a few months after Einstein published his theory in 1916.[5] He solved them for the special case of just such a collapse. He considered the example of a single point of great mass and how this point would distort the space and time in its neighborhood, and he found that both the point itself and the spherical region surrounding the point had strange properties. The radius of that sphere is called the *Schwarzschild radius* after its discoverer.

Consider a traveler making her way to the mass-point of one of these collapsed stars. According to anyone who happened to be watching the journey from a distance, it would appear to take an impossibly infinite amount of time for the traveler to reach even the surface of Schwarzschild's sphere. She would appear to slow down and after a while would look like a fly swatted against the sphere's surface. Even stranger, the light emitted by the traveler would also begin to fade, and any electromagnetic signals she emitted would drift to slower and slower frequencies, well below any receiver's ability to pick them up. Hence she would seem to disappear.

The traveler herself would not experience anything like this time distortion, but it would not be a comfortable trip. As she approached the surface, the force of gravity would begin to tear her apart because of what we call "tidal forces"—the same type of forces that cause ocean tides on the earth.[6] That is, suppose the traveler approaches the surface feet first. The gravity force acting at her feet would be enormously stronger than the gravity force at her head, and that force difference would tear her feet from her head—literally pulling her apart.

Furthermore, she would approach the surface of the sphere taking a normal amount of time, but then she would be pulled irrevocably into the sphere—retreat would not be possible. Not only would she be trapped in the region, but any light or electromagnetic signal she wished to emit would also be pulled back into the sphere, making her plight not only unremitting but incommunicable to others.

The Schwarzschild sphere, of course, is another name for the black hole—"black" because no light can escape from it, and "hole" because, as with any extremely deep well, once you fall into it escape is impossible. Interestingly enough, *Schwarzschild* means "black shield" in German.

BLACK HOLES MAKE WORMHOLES

Because the general theory of relativity foresees that space and time can be transformed by gravity, it makes some strange predictions. Further study of Schwarzschild's solutions for Einstein's field equations showed that the region inside the sphere would have a surprisingly complex structure. It could be a gateway or portal to another region of space—a hole in the interior of the sphere.

However, if you tried to enter the black hole at a slower-than-light speed, you would find the hole closing off as you traveled into it, and you would travel right to the center of the sphere where all of the mass existed as a single point. Any attempt to get near this mass-point, its *singularity*, would evoke even more horrific tidal forces than at the surface of black hole.

But if you were to travel faster than lightspeed, you could enter the black hole, travel through it unharmed, and find yourself possibly in some remote region of space far from where you entered. For any object going faster than light, the black hole would appear to be a kind of "wormhole" under space.

Schwarzschild black holes are now known as static, or untraversable, wormholes. A wormhole means a sub-spacetime tunnel that allows a traveler to move instantly from one mouth of the tunnel to the other, provided he could travel faster than light. Since gravity seems to choke off any possible travel, the question of whether or not they are "real" seems moot.

But suppose they were traversable, what would that mean? It would mean something could journey safely from one region of space to another via a black hole. Could traversable black holes be created and passed through safely? If so, how would we do that? It turns out that the answer depends on how strongly gravity stretches spacetime—in other words, it has to do with how much energy creates stress in a black hole.

THE SPEEDY SPACETIME TAILOR

Energy, although invisible, has the ability to thicken or loosen up, becoming denser or thinner, like the difference between honey when it is cool and gooey and when it is heated up and pours easily. There are, right now, invisible energy fields all around you, at your head and feet and thousands of feet above you, as well as miles beneath the earth's surface. Some of these fields we are familiar with, such as the electromagnetic fields responsible for radio, television, and microwaves. The microwaves around your head are most likely coming from radar and other detecting waves. Since they are rather weak in energy, they don't do too much harm. Inside your microwave oven is an entirely different situation. These waves are concentrated and meant to do harm—to cook your dinner. That difference between kinds of microwaves depends on the energy density—how thick the space is with microwave radiation.

For a black hole to become a traversable wormhole, it has to stay open for objects that move slower than light. The problem is that gravitational forces tend to close down black holes. One could say black holes are folds in the fabric of spacetime, and gravity acts like a speedy little tailor rushing to iron out those folds. To keep the wormhole open, something has to be inserted into the black hole to fight off the gravity force tending to shut it down. Since gravity produces positive energy, it makes sense that something with *negative* energy might hold the hole open.

Gravity acts attractively; it sucks stuff together like a thirsty kid at one end of a big soda straw. Negative energy would act in the opposite way, like a mischievous kid blowing air through a straw and making bubbles in his malted milk; it would blow the hole open. Hence the tailor could be "blown away" if there were enough negative energy causing stress inside the black hole, keeping it wide open and safe for travel.

Thorne's Traversable Wormhole

The traversable wormhole was used as a fictional space-transport device by the late Carl Sagan in his 1985 novel *Contact.* Prompted by Sagan, and on a bet with Stephen Hawking, Kip S. Thorne and his coworkers at the California Institute of Technology set out to determine whether traversable wormholes were consistent with known physics. So they re-solved Einstein's equations and came up with a solution that did not have a Schwarzschild radius where time and space went bonkers; nor did it have any excessive tidal forces to deal with. Hence Thorne's solution made it safe for humans to travel through a wormhole—but with one unfortunate drawback. Thorne's wormhole had to be threaded by exotic matter—the kind with a *negative* energy density, whereas ordinary matter has a positive energy density. If negative matter could be found somewhere in the universe, a negotiable wormhole would be a possibility. While a black hole only offers a one-way journey to nowhere, a wormhole would have an exit as well as an entrance.

You can imagine a wormhole as a tube that enters a subspace region under spacetime and provides the shortest pathway from one region of space to another, much as a subway tunnel passing through a mountain connects one part of the city to another via the shortest pathway available.

However, calling a wormhole the shortest-distance pathway doesn't really do it justice. For when you enter it at, say, point A and emerge from it several lightyears away at point B, you emerge at exactly the same time as you enter. Even though the entrance and the exit are separated by vast distances, you could walk through one mouth of the tunnel and emerge from the other at the same time![7] Crossing the galaxy would feel like nothing more than stepping over the threshold from one room to another.

Wormhole Time Machines

While science fiction has been using wormholes as speedy travel devices across space for some time, it is only recently that a new use for them came into view—as a means to travel across time, that is, as a time machine.

Thorne and his associates realized that one mouth of the wormhole could be moved in such a manner to put it out of temporal sync with the other mouth, thus allowing a traveler to enter the wormhole at one time and emerge at a different time. This isn't surprising if the traveler emerges at a future time—we all do that when we ride a subway train in one of the world's major cities. But in this case, the new time could just as easily be in the past! Since this movable mouth could then be placed close to the stationary one, the journey would be one in time but not in space.

Having arrived in the past, you could reenter the wormhole and return to the future where you came from. The only catch is that the "temporal subway" system has to be in place to begin with; that is, you can't go back in time any farther than the time when the wormhole was created.

Let's see how Thorne's wormhole would work as a time machine. Thorne described a scenario somewhat like this;[8] I've changed the story a bit to make it more interesting. Suppose that two long-lost lovers who have spent countless previous lifetimes together finally find each other in this lifetime. The only hitch is that he is thirty years old and she is barely ten! They of course would like to be the same age, but fate has dealt this cruel blow of a twenty-year age difference. So what to do?

Well, they happen to know Caltech professor Thorne and the fact that he has invented a time machine of sorts. They enter Thorne's laboratory to find, courtesy of Caltech's high-energy physics department, a complete facility for manufacturing three-foot-wide wormholes separated by twenty feet or so. One mouth of the wormhole, mouth A, rests in room A of the lab, and the other, mouth B, rests in adjacent room B. The man goes to room A and the little girl to room B. He puts his hand into the mouth of the wormhole in room A, and she puts her hand in mouth B in room B. (See figure 5.3.) Though they are on opposite sides of the solid wall separating the rooms, they can hold hands comfortably and even peek through the wormhole mouths to see each other in their respective rooms.

Caltech also has a special space travel facility right in the laboratory, and while the girl waits in room B, the man goes off on a rocket ship at near lightspeed, telling her he'll be gone for six hours and taking the mouth of his wormhole with him. Even though the mouths of the wormholes are now separated by a vast distance, the two can still see each other through their respective

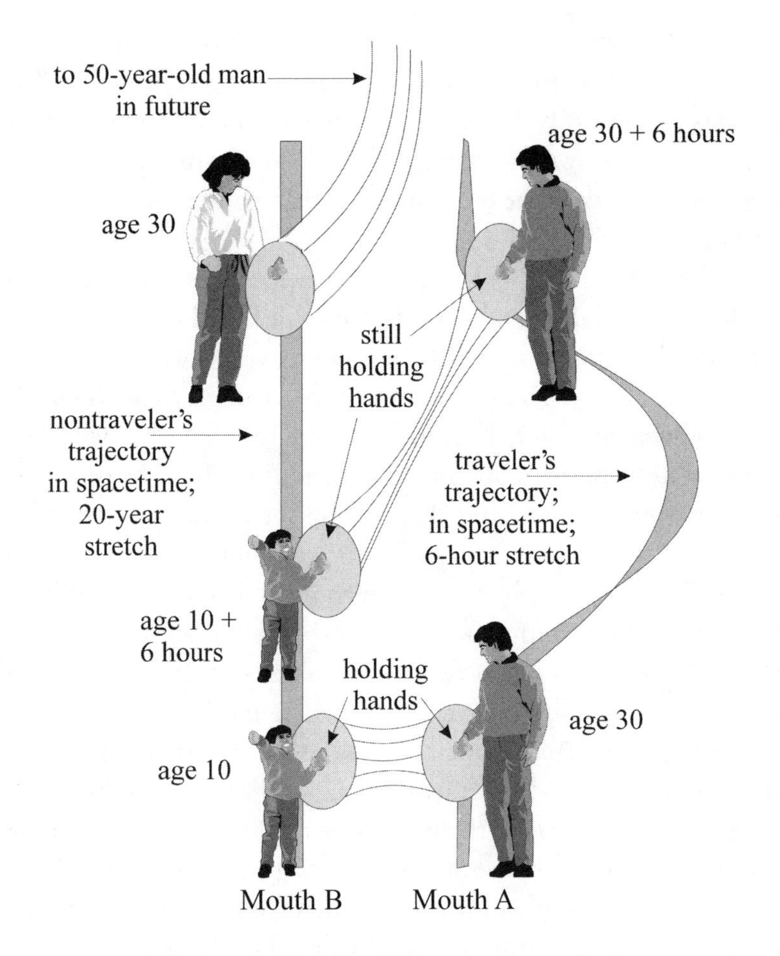

Figure 5.3. *A wormhole as a time machine.*

wormhole mouths and they can keep holding hands. As he travels, she can see the vast vistas of space he is traversing through her portal, and he can still see her sitting in the lab room B, and all the while they are still holding hands.

Checking the clock aboard the space ship, we see that after about three hours the ship turns around and heads back to earth, making the round trip in about six hours. When the ship lands, the man takes the mouth of the wormhole with him and returns to room A at Cal Tech where he started, all the while still holding hands with the little girl in room B.

He then walks over to the door to room B, opens the door, and finds a lovely thiry-year-old woman smiling at him. The little girl has grown up while she was waiting for him during what she experienced as twenty years at Caltech—even though he is still holding hands with the ten-year-old girl she once was through his end of the wormhole.

Meanwhile, from her standpoint, she counts the hours until six have passed. She looks through mouth B and sees that he has landed! She watches him through mouth B as he walks into room A. Excited to greet him, she runs to the door to room A, all the while giving his hand a gentle squeeze through the wormhole opening as he reassuredly squeezes back. But when she opens the door to room A, he is nowhere to be seen. She looks into the room and she is alone. She looks through mouth B and sees him smiling back at her while he sits in room A, his other hand holding the hand of a rather attractive and familiar woman. "Who is she?" she wonders jealously. And the older woman speaks to her through the wormhole, saying, "Hi, young me!"

Bizarre as this may seem, as far as the little girl is concerned, the man has not returned to planet Earth yet and won't do so for another twenty years, for now the two times are out of sync. Even more bizarre, she is in two times and thus at two ages at the same time.

As in the example of the muons living nine times longer than their usual lifetimes discussed in the opening chapter, the relativity of time dilation has taken its toll. Even though he himself has only spent six hours making the round trip, from her perspective he won't reach earth for another twenty years. Yet she decides to wait, and except for brief moments of sleep and food intake and the like, she maintains her hold of his hand. As the next twenty years go by, she notices his hand is aging and wonders just what he will look like when she reaches her thirtieth birthday.

Once those twenty years have past, she watches as his spaceship finally lands and he walks out of the ship looking no more than six hours older, while she has aged twenty years. He carries his end of the wormhole, mouth A, with him, through which he is still holding hands with her as a ten-year-old child. She, having become a young woman, wonders whose hand she is holding through mouth B of the wormhole. She knows it is her love's hand, but she notes it is the hand of a fifty-year-old man, not the thirty-year-old she sees before her and remembers from when she was ten.

They sit down together in the same room; he now admiring the thirty-year-old woman she has become, and she realizing that now they are the same age and can get married. But, what about the wormhole and all of that hand holding? She looks through his end of the wormhole, mouth A, and, much to her surprise, sees herself as she was twenty years earlier, a ten-year-old still holding his hand. She looks through her own end of the wormhole, mouth B, and sees her love not as he is now but as he will be twenty years hence, a fifty-year-old man. If the fifty-year-old man likes, he can squeeze himself into mouth A of the wormhole and emerge twenty years in the past to find his sweetheart just as she was at age thirty. She can enter her end of the wormhole, mouth B, and travel twenty years into the future to find her fifty-year-old man.

What a relationship! By moving forward and backward through their respective wormhole mouths, they can meet at a whole range of ages. She at the age of ten could squeeze through her end and travel twenty years into the future and meet herself at the age of thirty. Her ten-year-old self could then squeeze through mouth B of the wormhole again and emerge another twenty years into the future to find her man old enough to be her grandfather. Talk about May-December relationships, theirs is a whole calendar's worth!

What has happened? His end of the wormhole is now twenty years in the future from the time she was just ten, and her end is back in time twenty years from their current thirty-year-old equal-age time. From here on out, the two wormhole mouths move on through time at the same rate, so that a year later he can return to the past to find his sweetheart as an eleven-year-old, and so on. The wormhole has now become a full-fledged twenty-year time travel device. By moving the ends again, it would be possible to extend that time, but traveling backward through time to any time earlier than the wormhole's first construction would still not be possible.

CHAPTER SIX

The PARADOXES *of* PHYSICAL TIME TRAVEL

Time is a great teacher,
but unfortunately
it kills all its pupils.
—Hector Berlioz

The general theory of relativity predicts time travel, but there are still problems to consider, specifically, the paradoxes that time travel introduces. In this chapter we will explore whether it's possible to get around these paradoxes, or whether they are fatal, driving the final nail into the coffin of time travel as a reality.

It is widely believed that time travel is impossible because it violates the laws of physics. In fact, skeptics and others who have contemplated the logical paradoxes of time travel figure that a physics theory predicting time travel in itself shows that theory to be false. Is this actually the case? Do the laws of physics prevent time travel? The answer turns out to be a surprising and resounding no. Not only do the laws of physics *not* forbid time travel, they may require it!

CHAPTER SIX

THE CREATIVITY PARADOX

First we'll look at a chicken-and-egg paradox that would arise with the invention of time machines. Oxford University philosopher Michael Dummett modified the age-old poultry question into what is called the creativity paradox or the knowledge paradox.[1]

Consider the story of a very creative scientist who comes up with a new theory in the year 2100, just as she turns fifty years of age. Since it is a receptive moment for new ideas in science, the theory is welcomed and acclaimed as brilliant; it makes possible, among other novel technologies, time machines. With funding from her local university, she constructs a time machine. Then she decides to use it to send her book and all of her notes on how the theory about time travel works back in time to her younger self when she was just a baby. Accompanying the book is a note saying: "Do not open until your thirtieth birthday." Her younger self, who earned her doctorate in physics at the age of twenty-eight, opens the book on her thirtieth birthday and discovers that it is a complete text on the theory and construction of time machines. But since science in 2080 has reached a conservative nadir when new ideas are quickly scoffed and discouraged, she decides to wait and publish her discovery later. Somehow, she finds herself intuiting that the year 2100 is the time when these ideas will be welcomed, and sure enough, when she turns fifty that year, she copies down the theory exactly and publishes it as her own work. Based on the book, she builds a time machine.

Then she sends a copy of the book back in time to her younger self as a baby, with a note saying not to open the book until the child's thirtieth birthday . . . and the story continues in this bizarre time travel cycle.

The story seems to suffer the fatal flaw of circular reasoning. However, it actually is not a vicious circle, which assumes the conclusion in the premise. More to the point, regardless of its seeming nonsensical twist, the story does not violate any law of physics.

Nevertheless, something is clearly wrong. If the copy of the book is the source of the book, then when and how was it first created? As we shall see shortly, the question that can be answered will be: Where was it created?

The Grandfather Paradox

Science fictions fans are familiar with another time-travel puzzle known as the grandfather paradox. In the standard literature, the time traveler goes back in time and kills her grandfather when he is still a boy. So how, then, could she ever be born? Here, I've changed the characters a bit, but you may recognize the plot. A bright young scientist who has a particularly creative and imaginative relationship with her mother invents a time machine and goes back in time to visit her mother just before she marries her father. She tells her young mom-to-be who she is, and her "mom," being very imaginative herself, becomes quite excited. "Mom" then goes out on a date with dad-to-be. But when she tells him this "far-out" story, "dad" doesn't believe her and thinks she is nuts. He decides not to marry her, and the bright young scientist never is born.

If we believe this story, time machines are impossible because they would lead to a logical inconsistency: If a future action A leads to a consequence (action B) in the past that prevents that action A from taking place, then how could that action A ever occur to begin with? In this case, how could the bright young scientist return to the past (action A) and prevent her parents from getting together to have a child (action B), when that action B would in turn prevent (action A) from ever happening—since she would never have existed to travel back in time? Hence this *grandfather paradox* is also called the *consistency paradox*. I used a variation of this story in my book *Parallel Universes*[2] to explain how classical physics, which never brings such paradoxes into account, could not handle them even if it did.

Even though classical physics cannot handle time travel, in fact nothing in its structure forbids it. The reason has more to do with the logical structure of science, and physics in particular, than it does with the laws of classical science. Logical structures in science are the same as logical structures in any field of enquiry. They refer to such things as the familiar cause-effect relation: if A, then B, and so on. Laws of classical science deal with the specific content contained within these logical structures. For example, in physics imagine a ball (mass A) at rest on a shelf, falling off into a gravitational field (force). It will thus accelerate downward. We could write the classical physics second law of Newton that governs this example: If a force is applied to a mass A, then mass A will accelerate. Now freeze the frame, so to speak, run the film backwards, and imagine the ball reversing its motion but having the same speed moving upward against the gravitational field, finally slowing down as it comes to rest on the shelf. Classical physics really has no problem with events appearing in a reversed time order. Its laws are what is called *time-reversal invariant,* which means that if the objects being observed were to reverse their motions because we began to count time as going backward, nothing in those laws would need changing.

So where does the paradox come in? It arises because classical physics insists by fiat that time and space are immutable—they are not be messed with. Once an object follows a path from here to there and does so in a certain amount of time, the deed is done, the die is cast, and nothing can wipe out or modify that history. Although many people think that this is a rule of physics, it actually isn't.

The Chronology Tenet

Up to now I have tended to weigh the words *science* and *physics* equally, as if they were one and the same. Certainly, *science* is the more general term and *physics*, both classical and quantum, fits within science's defining boundaries. Most of today's sciences undergo change as the discoveries made in them through the use of quantum physical principles become more evident. Even though it may seem to the nonscientist that quantum physics evades logical description, the rules of logic are firmly present in it and in all fields of science. Consequently, all of these fields base their logic on common sense.

Classical physics is based on a number of commonsense principles like the one we have just considered: the immutability of time and space. They are often hidden from logical view and remain scientifically unprovable, yet they play useful roles. One such principle is called the *autonomy principle.* It says that we can do experiments in our local neighborhood without reference to or concern for the rest of the universe. For example, I can open the tap on my water faucet, fill my teakettle, and boil water on my stove and expect that my actions will not be inconsistent with the behavior of the rest of the universe. Said differently, I can assume that the actions of distant stars and planets need not be taken into account when I boil water for tea or decide to take the dog for a walk. If this principle wasn't true, any action I take would evoke all kinds of consequences for the rest of the world or even the universe at large.

Magical belief systems base their logic on just such notions: that we are not autonomous, that everything is connected. Hence a tribal shaman would perform a rain dance based on the belief that his action would cause a change in the sky. Or someone who is ill might ask God for healing, based on the premise that mental action causes a response from larger forces, or rub a rabbit's foot with the hope that this shifts the lines of fate to bring good luck in the future.

Together, the autonomy principle and the grandfather and creativity paradoxes constitute an unstated assumption called the *chronology tenet* that concludes simply: You cannot move backward through time. If the chronology tenet is not actually a part of physics, why do we believe in it? The reason has more to do with our commonsense view of causality than with physics. We believe that if something happens, there must have been a prior cause plus a means by which the cause leads to the effect. Travel into the future does not violate this commonsense view, but travel to the past, even for the briefest of times, apparently does.

Let's examine this conclusion more clearly. We have seen how the general theory and the special theories of relativity introduce new ideas about time. Specifically, through the time dilation effect,[3] which briefly says that when in motion time slows down, the special theory of relativity allows us to travel to the distant future while aging minimally. Here no causality paradoxes are encountered, even though such travel would certainly be weird.

The general theory of relativity lets us construct wormhole time machines that do allow travel to the past, but no further than when the wormhole's "past" mouth was discovered or first created. Here we *do* encounter causality paradoxes, so it seems that travel to the past is the more troublesome. But if we consider carefully, we see that the chronology tenet is specifically relevant to matters of cause and effect, not necessarily to traveling in time. It says, in essence, "Thou shall not violate the laws of causality."

But does travel to the past really violate the laws of causality? If we examine closely what we mean by time in light of quantum physics principles, we find that travel to the past does not violate any of the paradoxes or principles of the chronology tenet, and in fact remains within the laws of causality.[4] To see how this is so, we need to consider how the physics of general relativity predicts the existence of what are called *closed timelike lines.* These are trajectories through space that at first move forward in time but then curve around and go backward through time, arriving right back where they started at precisely the time they started. The word *timelike* here just means that any motion along that line will appear to be going in the direction of ever-increasing time for the one moving along it less than the speed of light. The lines are called *timelike* because anyone moving along them would experience the clock ticking along normally. When the person arrives in the past, the clock would show a normal time advance. And the lines are called *closed* to remind us that they loop backward in time. Later, we will see what happens when we open them, but allow them still to go back in time. These closed lines in spacetime were first found to be solutions to Einstein's general theory of relativity equations by Kurt Gödel.[5] Although they seem to be physically impossible, it turns out that the general theory of relativity does not find them so.

We also need to examine quantum physics' notion of parallel universes: the proposition that instead of a single universe containing all there is, there are an infinite number of universes; the matter in all the other universes manifests as "parallel ghosts" when contemplated from any single universe.[6] All of time travel's paradoxes are resolved, provided that closed timelike lines can be opened so that they thread their way into parallel worlds.

Parallel Universes

Australian aborigines refer to the concept of parallel universes when dealing with Dreamtime—the sacred law governing their existence. Each instant of normal time "contains," as if in a parallel world, all of the sacred history and expectations of the people. Even the land "contains" this history and expectations. Each instant of time contains a sacred hoop of Dreamtime or spiritual history that repeats itself.

Even though quantum physics and Dreamtime appear quite unrelated, since one arises from objective science and the other from subjective spirituality, there may indeed be some overlap. I can only speculate here, of course. Dreamtime refers to an ancient period long before humans appeared on the planet. In the theory of parallel universes, particularly as David Deutsch thinks about them, past times and future times are just parallel universes.[7] Hence, Dreamtime is a parallel universe of a past time that from instance to instance may overlap with this one.

Parallel universes have existed in the fantasies of science fiction probably ever since the genre began. Even though they are a subject of debate in science proper, science fiction writers apparently have no problem dealing with them, and readers have come to accept them. One of the best parallel universe stories is Sir Fred Hoyle's science fiction masterpiece, *October the First is Too Late.*[8]

Hoyle, who died in August 2001, was one of the most visionary scientists and was largely responsible for discovering how the elements from lithium to iron are synthesized inside stars. Despite the fact that he coined the term "Big Bang" to describe the theory that the cosmos was created by a huge explosion 15 billion or so years ago, Hoyle didn't accept this theory. Instead, he advocated that the universe has no beginning, and that new galaxies form in the gaps created as other galaxies move apart. In spite of observational evidence to the contrary, he continued to attack the Big Bang theory throughout his life.

In his story, set in the year 1960, whole populations on the earth, with the exception of England, are replaced by a bizarre quiltwork of past and future civilizations. Modern Greece reverts to the Golden Age of Pericles, while Russia and Asia are thrown into a far-distant future witnessed by our then dying sun perhaps billions of years from now when life on that continent cannot be sustained. America is thousands of years into the future with increased technological advances. Hoyle introduces the fantastic idea that the world exists this way because of the abilities of a godlike super-mind to conceptualize it in this manner. The future worlds and the present world exist side by side, due to the observational efforts of this mind. Eventually the different time periods return to "normal," when past and future times are not mixed into present times.

Hoyle based his tale on the parallel-universes interpretation of quantum physics by implying that other times could coexist as if they were parallel universes overlapping with ours in our time period. It may appear strange to the reader that a theory of science like quantum physics can have different interpretations. Although quantum physics is a well-founded theory and its equations are certainly not in question, there are several competing understandings of just what quantum physics means. Surprising and fantastic as it may seem, the parallel-universes interpretation appears to be the most straightforward, but certainly controversial, understanding of quantum physics to date. However, regardless of interpretation, quantum physics has proved itself to be the best physical theory we have to date. So far, nothing contradicts its bizarre predictions, and we have been witnessing its implications for well over a hundred years. In fact, it is the basis of today's age of information technology.

How did this parallel-universes interpretation arise? In 1957, the late Hugh Everett III, then a graduate student at Princeton University studying under the highly regarded physicist John Archibald Wheeler, came up with the rather strange notion that we should take quantum mechanics (which is the same thing as quantum physics) seriously.[9] Everett noted that quantum physics predicts that all alternative outcomes of any given experiment must occur even though we may only see a single outcome! Somehow, those hidden alternatives must exist simultaneously along with the observed outcome.

In contrast to classical physics in which a single outcome of any experiment is determined by the implicit laws of causality implied, *all* outcomes of any experiment are predicted to actually occur in the parallel-universes interpretation of quantum physics—each weighted by a probability.[10] Having multiple outcome possibilities is indeed an important feature of quantum physics, according to most physicists' view of the quantum theory, provided nothing is observed prior to any outcome. It's just what happens to these alternatives after an *observed* outcome that remains in question.

Suppose, for example, you have a single die that is weighted so that the number 3 arises 50 percent of the time. The other five numbers are weighted equally—each with a probability of 10 percent. In a classical worldview, if we roll the die enough times the results will conform to this probability distribution—the 3s showing up half the time, and the other five numbers each showing up 10 percent of the time.

But in the parallel-universe view of quantum physics, when the same weighted die is rolled, all of the numbers show up after a single roll. The reason we see only one number is that each time an observation occurs, the observer splits and enters into each of the six worlds predicted—again her appearance in these alternate universes being dictated by the weighting of the odds. The world where she observes the number 3 is five times as likely as any of the other worlds where the other numbers are observed by parallel "shes." Yet all of the worlds supposedly occur. How are we to understand one world being "heavier," in a probability sort of way, than any of the others? This question, in fact, is what convinces many physicists that parallel universes are implausible.

However, let me speculate. One possible meaning is that the other five worlds combine to form a single world. The consciousness of the observer splits equally between the two worlds—the world of number 3 and the world of all of the other numbers combined. In the one world, he knows the number is 3; in the other world, he knows the number is not 3, but he doesn't know which of the remaining five numbers actually shows. To resolve that probability, the other world now splits into five non-3-showing worlds of equal weight. In each of these none-3 worlds the observer's consciousness knows both that the number is not 3 and which of the new numbers it is. In my addition to the usual parallel universes idea, all outcomes must be equally weighted off any split so the question of unequal weights becomes moot. The first split was equally 50 percent to each world. In the second split, all five worlds split off with equal weights of 20 percent (as far as the original non-3 world is concerned) to each world.

In spite of its bizarreness, the parallel-universes view of quantum physics remains a completely deterministic theory—it accounts for our subjective experiences by providing a reasonable history of possible outcomes we would, or could, see in each of the worlds we happen to inhabit. In other words, in whichever world we are, the results we see will be consistent with the classical view. It explains probability outcomes by taking all outcomes into account, provided we let the observers of those outcomes multiply without end.

Any universe you may inhabit at the moment will seem real enough with the others hidden from plain view. However, the same thing will be true for each of the other universes and other "yous" as well. That's what makes parallel universes seem so unbelievable; how can there be copies of me that I have no knowledge of?

What Happens in a Parallel Universe?

Parallel-universe theory says that a universe can have parallel nearly identical copies of itself, and no one would ever know it. In effect, the universe you are in can be splitting into nearly identical copies all the time, every time someone observes something. Every time you make an observation, the world splits into as many possible outcomes as you could witness from your one observation. For instance, when you flip a coin and see it land, the world splits in two—a "heads" world and a "tails" world. What is even more unsettling, the mere act of watching the coin land splits you into its two worlds as well. In each of these worlds, each "you" sees the coin with its appropriate side showing. The "heads-you" sees the coin facing heads up, and, in the other world, the "tails-you" sees it in the opposite way. But if that observation was made by another person, then these other universes you unwittingly happen to be a part of will be identical to you. As far as you are concerned, these other worlds overlap, and there is no way for you to realize they exist. This may appear quite bizarre. You can think of it in two ways. You exist in parallel worlds that are indistinguishable from each other as far as you can determine. Hence, they don't exist as separate worlds for you. For you and your "twins" there will be just one "you" experiencing one universe. Thus we have the remarkable situation that one universe at any time has an infinite number of identical copies, and you are in all of them doing exactly the same thing you are doing in this universe—wherever and whenever "this" universe happens to be.

I suggest that this "fact" of quantum physics lies at the root of the inherent stability of all atoms and thus all material objects, and that it is largely responsible for the inertia of objects and why matter appears as solid as it does. Every object in the universe, large or small, is incessantly splitting, and these splits, resulting in multiple appearances in multiple universes, reinforce each other, making the objects appear solid in every one of the universes. To illustrate, think of a rapidly rotating two-blade airplane propeller. Those two blades trace out a convincing disk as the propeller rotates, and for any who get too near it, the reality and solidity of the disk is dangerously real.

If you can't see any difference between any of these worlds, then why bother with them at all? That's where the closed timelike lines come in. We need these parallel universes to make time travel a reality, and we need closed timelike lines to make the paradoxes of time travel go away. Closed timelike lines must open out and thread their way through these universes in order that no paradoxes occur.

Remember the example of the four-dimensional tube discussed in the previous chapter? That tube stretched across time from birth to death and, in terms of timelike lines, is an open tube of lines. But just imagine the death end of the tube joined to the birth end, so that the tube now becomes made up of closed timelike lines. The paradox immediately arises that the person is now not only in two places at the same time, but is also two people of different ages at the same time!

As we see in the upcoming resolution of the chronology tenet, time-travel paradoxes are resolved by opening closed timelike loops so that they thread from one parallel universe to another. Now, for example, when a dying traveler goes back in time to his birth, he ends up in a parallel universe where he sees his parallel self being born. Strange, yes, but not forbidden by the laws of physics.

RESOLVING THE PARADOXES IN THE CHRONOLOGY TENET

The presence of parallel universes and closed timelike lines actually helps resolve the paradoxes discussed above inherent in the chronology tenet. Consider the grandfather paradox—the example of the bright young scientist who goes back in time to talk her younger mom out of marrying her dad. If she can find and make use of a closed timelike line to go backward in time, then the instant she appears in the past, the universe splits into two copies that are nearly identical. Instead of unconnected parallel universes, each containing its own paradoxical closed timelike line and a copy of the time traveler, there are two parallel universes threaded by a single (now opened) closed timelike line. In universe A, mom marries dad; in universe B, the time traveler arrives in the past, and mom doesn't marry dad. (see figure 6.1.).

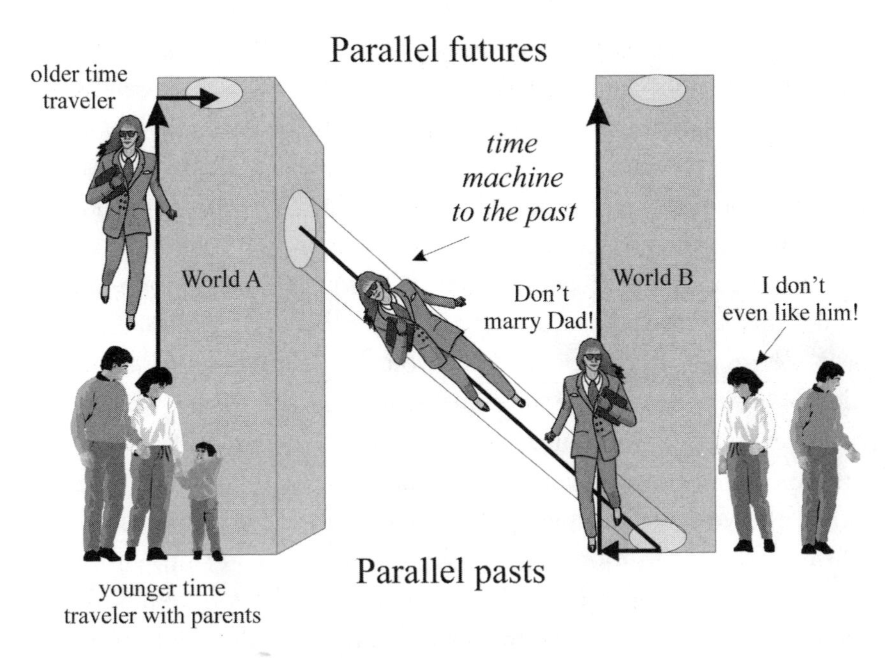

Figure 6.1. *The grandfather paradox resolved with a time machine.*

In other words, the time traveler leaves universe A and takes a trip to the past, arriving at universe B. Until she emerges in the past, A and B are the same. When she arrives in the past, universe B splits off from universe A. In universe B, mom does not marry dad and the daughter is not born. The paradox is resolved by the second universe. In universe A she is born and in B she is not. As far as residents of B are concerned, she is a visitor from another parallel universe.

Consider another paradox similar to this one, even though it may not appear to be so. Suppose a traveler decides to travel to the past. We can imagine the time machine with two portals—one in the past and one years ahead in the future. He plans to leave through the future portal in the year 2030 when he is thirty years old and emerge from the time machine in the year 2000—the year of his birth—through the past portal. He wonders what will happen if he sees himself emerge through the past portal. He is now twenty years old, and so far no one has emerged through the past time portal—yet. But he makes a pact with himself that he won't make the trip if he sees himself emerge in the past time portal. However, this leads to an interesting paradox. If he makes the trip, he will emerge in the year 2000 where he will see himself as a baby. When that baby grows up, the man the baby becomes will see him and therefore not make the trip. But if he doesn't make the trip, then he won't emerge in the past, and he will then make the trip according to his pact. How do we get out of this paradox?

If he enters the machine when he becomes a young man, he will emerge in the past and meet a younger (baby) version of himself. Consequently, his parallel and younger self will see him and decide not to enter the machine when he gets older. If this younger self does not enter the future portal, he will not emerge in the past and meet a younger version of himself, in which case when the younger version gets older he will enter the machine through the future portal, go back in time, and meet up with a younger version of himself. This "tempo-mental" monkey business reduces to: If he doesn't go back in time, he does, and if he does go back in time, he doesn't.

To make this idea clearer, consider Figure 6.2 where we see how to resolve the paradox using parallel universes and an "opened" closed timelike line threading the universes. You can see that the line is nothing more than a Thorne time machine, but this time connecting two worlds rather than two people in the same world (see chapter 5). Here we want to consider what happens in both worlds when our traveler leaves one and goes to another. Like the previous time traveler who makes an interworld journey, our traveler decides to enter the time machine in the future portal when he is older and emerge in the past. But here the pact is in effect.

Thus we have the paradox resolution as shown in Figure 6.2: All is resolved by the presence of two branches of parallel worlds, which we can label as worlds A and worlds B. Yes, I mean two distinct sets of worlds. In worlds A, he enters the machine and consequently travels to worlds B. In the B worlds, he meets with a baby version of himself, and his baby twin does not enter the machine when the twin reaches his present age. Any friends left

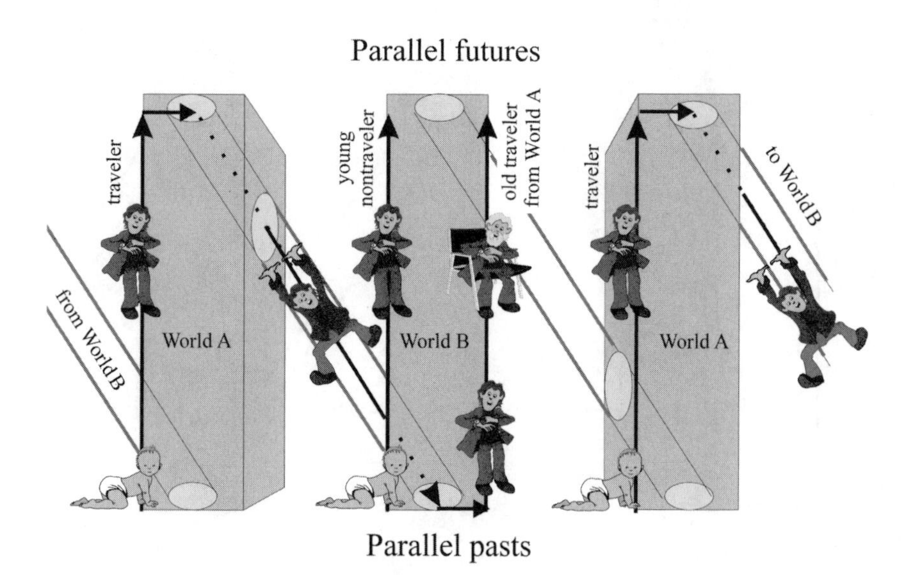

Figure 6.2. *Parallel-worlds spacetime blocks with a time machine.*

behind in worlds A will say he has vanished. Any friends of the younger version of himself in worlds B will say the B version now has an older brother (namely himself). The closed timelike line threads its way through all of the branches of parallel universes in such a manner as to keep things consistent.

Finally, let's look at the resolution of the creativity paradox mentioned above (see figure 6.3). Here, there is one original world, world A, where the manual is actually written and many parallel copy worlds, worlds B, where the manual is simply copied. We have again parallel universes threaded by a closed (but now opened) timelike line that originates on one world and then threads its way through several duplicate parallel worlds. In world A, the scientist creates the time travel manual and then sends it on its way through the future portal to the first parallel world B. In that world, the manual is copied and then sent on to the second, duplicate parallel world B, and so on. Here we have

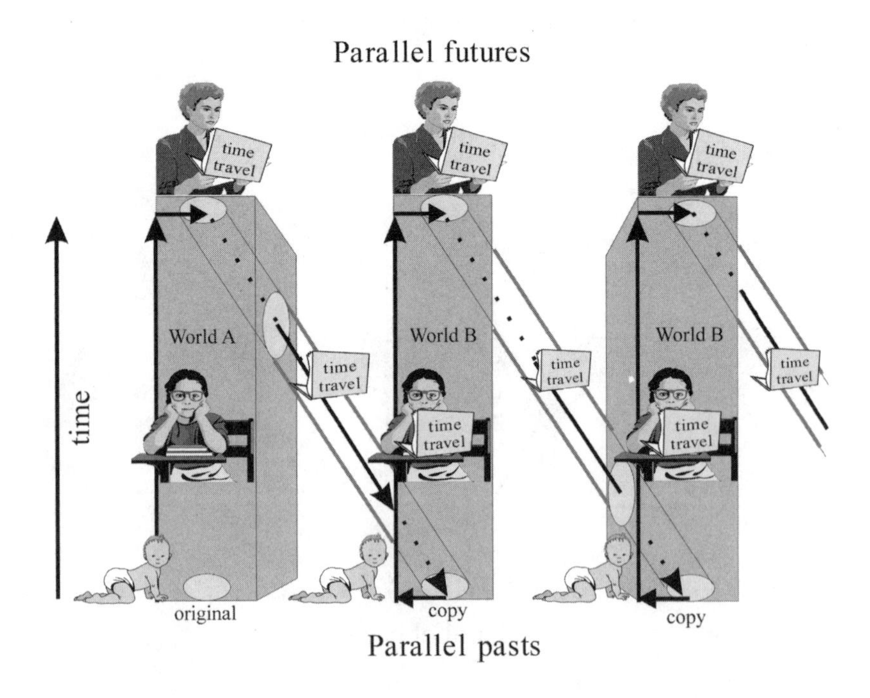

Figure 6.3. *The creativity paradox resolved by parallel universes.*

one original world and many identical copies, although none of the copies is really the same as the original.

All of the above paradoxes can be resolved by dividing the parallel universes into two (or more) branches threaded by a single closed timelike line, enabling the traveler to connect in a consistent manner all of the universes wherein no paradox arises in any single universe. Certainly the arrival of an older twin or a book in a universe will be surprising, as will the vanishing of either from another universe. However, as surprising as these events may be, they do not violate any law of physics.

CHAPTER 7 SEVEN

The TECHNOLOGY *of* "ORDINARY" TIME TRAVEL

Daylight-saving time is when the government tampers with God's time.

—Anonymous

By ordinary time travel, I mean travel using a contraption that literally conveys the time traveler from one time zone to another—for instance, from the present to the future or from the past to the present. We have seen imagined examples of such devices in science fiction movies, such as *The Time Machine* based on H. G. Wells's story, which has been made into a movie at least twice by Hollywood and, I would guess, may be made again as special effects technology continues to improve.

The film portrays what most of us think of when we ponder time travel: Our hero sits inside a machine and witnesses scenes passing by his eye very quickly—an effect achieved through time-lapse photography.

However, when we think of traveling in a time machine, we are not all that interested in experiencing the journey itself. We don't really want to stay in the time machine any longer than necessary. We want to reach a destination—some time period in the past or future—and then descend from the carriage, so to speak, and arrive on the scene.

In these fictional examples, the time traveler moves through time as if he were moving across space. He travels backward or forward in time just as you might imagine traveling back and forth from one city to another and back again on a bus, car, or airplane.

In this chapter, I want to explore a new device. In contrast to wormhole tubes discussed in the previous chapter, in which the traveler remains relatively unchanged while she journeys to the world of the future or the past, this device would enable the person to time travel while the rest of the world moved on in ordinary time. In brief, the traveler's knowledge, her physical state, and her perspective on reality all shift in time. Or to put it another way, while the world apparently goes on its merry way, the time traveler changes by gaining knowledge of the future or the past. When she leaves the device, the time traveler will also find herself younger or older than when she first entered it.

The key to making this time-travel device work is its connection to a special kind of computing device known as a *quantum computer.* A quantum computer produces a strange kind of "data processing," not the usual number or letter variety. Moreover, this data processing can't be observed by anyone without the data changing in an irreversible way—once it changes, you can't get it back again.

Unfortunately, this time machine won't always work when you want it to! In fact, as the designers themselves explain, it will work only rarely because of the probability nature inherent in quantum physics. But if and when it does work, all I can say is hold on to your hats, for the world will never be the same.

The quantum computer, a new type of computing machine that operates fully in accordance with the laws of quantum physics, is based on the ideas of physicist David Deutsch, who pioneered the theory in the 1980s and mid-1990s.[1] Quantum computers are now being researched in several laboratories worldwide. Although there has been quite a bit of experimental development, for the most part theory remains ahead of the actual devices, and it is probably safe to say that practical quantum computers in the home won't be a reality for another ten years. But who is to say when? While today's computer, utilizing microchip technology, also follows the same quantum laws, the computer's user doesn't access these laws directly. Like so many other functions in our modern computers, they are deeply hidden, far deeper than any programmer has access to. However, they are used daily by the makers of the microchips that form the data processors and memory in the "motherboards"—the main processing components.

Users of quantum computers, however, *will* have access to the bizarre quantum world that is hidden from the "classical" computer user. Just what this will mean to users when quantum computers become mainstream remains to be seen. Whatever it means, the world will be stranger indeed, and device-assisted time travel may indeed become a reality.

THE LAWS OF QUANTUM PHYSICS

Quantum physics, or quantum mechanics, which is the same thing, is a strange business. It deals with the behavior of matter and energy, particularly with how matter and energy interact on a very, very tiny scale—the scale of atoms, molecules, and the particles that exist inside these small objects.[2] It also covers the behavior of large objects discussed in previous chapters, as I'll explain shortly. On this tiny stage, the miniscule atomic and subatomic actors do not behave as their constituencies do any more than you behave as your country does. Just as the laws of a country are based on rules that may not apply to an individual (for example, your country can lawfully print money while you certainly cannot), the laws of quantum physics governing the behavior of subatomic, atomic, and molecular objects are different than laws governing large objects such as human bodies, stars, and rolling dice.

Yet we all know that a country's behavior is to some extent based on the laws an individual follows. Similarly, the laws of quantum physics governing individual tiny-scale particles determine the behavior of the larger objects made from these tiny particles. This understanding of large-scale behavior comes about by determining the average of many small-scale events. For example, insurance companies determine your premium rates based on the average age and health of individuals in your city, state, or country.

Until very recent times, it was believed that quantum physics only applied to the atomic and subatomic world, a world that was well below human perception. Today, scientists believe that quantum physical effects can also be observed on a larger time and space scale, well within the world of human perception. However, in contrast to large-scale movement, where Newtonian or classical laws of motion apply, quantum physics laws do not determine ahead of time what will actually happen in any given situation. Instead, much as statistical laws are the basis for constructing actuarial tables, quantum physics laws let us calculate very accurately the probabilities for events to occur, even while we remain completely in the dark about the actual events themselves.

The situation is even stranger than one might imagine. Classical physics deals with numerical probabilities all the time. Whenever we can't calculate the outcome of an experiment because of the impossibility or inherent difficulty to control it, we rely on probabilities. For example, we normally can't control whether a flipped coin will land heads or not. But quantum physics works quite differently. To arrive at a probability for a sequence of events, you have to imagine the possibility moving as a wave through time from a specific starting point (for example, the flipping of the coin), then reversing itself when the wave reaches a specified future time (the coin landing on the floor), and finally coming backward through time to where it all started. These two "flows" of *possibility*-waves then come together multiplying each other.[3] I will say more about these strange, time-reversing waves in chapter 8.

Possibilities can also overlap and add together. The overlap and addition of two or more possibilities is called a *superposition.* If you were to imagine each possibility as possible routes between Chicago and New York, each drawn on a separate sheet of clear plastic, superposition occurs when you put the drawings on top of each other so that you can see all the routes at the same time.

Superpositions of possibilities can produce curious results. For example, the side of a flipped coin is, in principle, predictable in classical physics if one had the ability to control all of the variables involved in the actual flipping. Bearing with our lack of control, one usually assigns a probability of 50 percent that it will land heads and 50 percent that it will land tails. However, in quantum physics these separate possibilities can *superpose,* leading to new possibilities—for instance, the coin landing standing on its edge.

When a *possibility*-wave completes its turn-around cycle and multiplies with itself, the possibility becomes a probability. Physicists now believe that at this point, the event in question is "waiting to be observed," so to speak; even though it has not yet been observed, it can no longer be termed unobserved. In fact, usually, the completion of a cycle and the final observation of an event are simultaneous. Hence, the coin, which was previously capable of existing in one of two possible states (heads or tails), suddenly jumps into one of those states (say, tails) at the instant it is observed. This is called the *quantum physical observer effect.*

When Two Possibilities Make No Possibility

The classic example of the superposition of possibilities and the observer effect is the famous double-slit experiment. In this experiment, a stream of subatomic particles is directed through a screen containing two very closely spaced narrow slits (like parallel Venetian blinds). The setup allows particles to pass only one by one through the slits. Each particle makes its way to a second screen, where it hits and makes a single tiny spot (see figure 7.1). One would expect that each particle in the beam must pass through either one slit or the other in order to reach the screen. Yet after many particles have made their way through the slits, a pattern of dots appears on the second screen that can only be explained if each particle somehow passed through *both* slits simultaneously. In other words, the two possibilities—one slit or the other—seem to superpose to produce a new result.

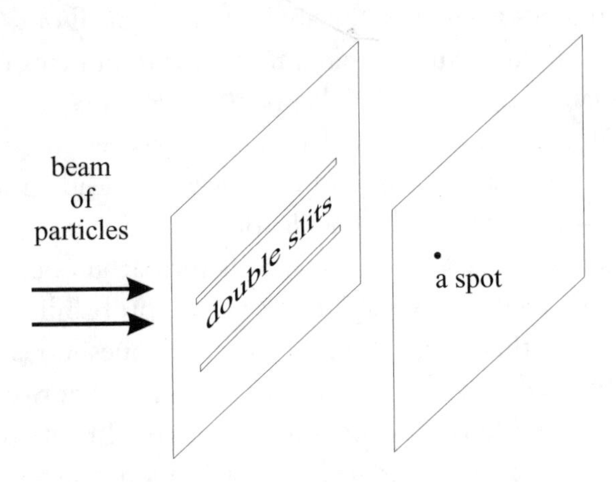

Figure 7.1. *The double-slit experiment.*

Amazingly, if you close down one of the slits, more particles reach certain places on the screen than if you leave both slits open. Look at figures 7.2 and 7.3. Notice that, for example, with one slit open the particles fill in a wide area, while with two splits open there are gaps showing up in the same area.

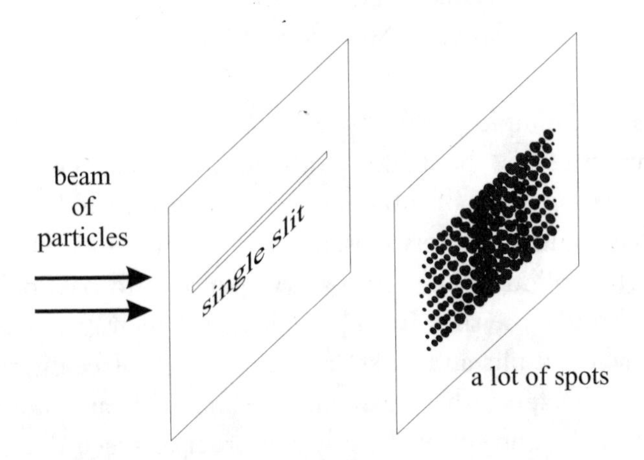

Figure 7.2. *A single slit makes a lot of spots.*

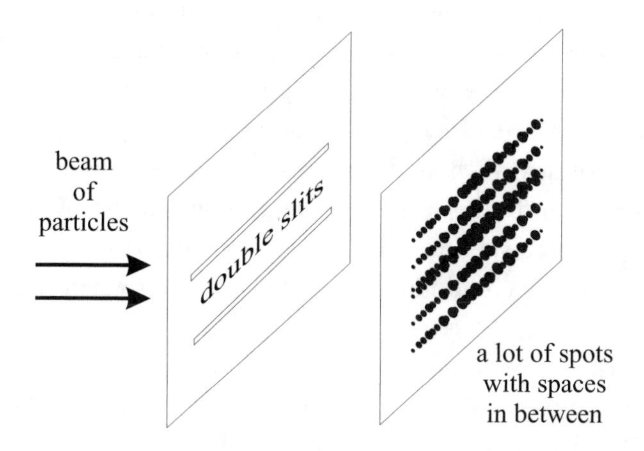

Figure 7.3. *A double slit leaves white spaces in between bars of spots.*

There is really no way to understand this fact if you think that the stream is composed of single tiny particles. With both slits open, each particle apparently has two choices of which slit to pass through, so each should have twice the opportunity of reaching any point on the screen. In other words, any point on the screen should be twice as likely to record a hit, so there should be fewer open spaces between spots when both slits are open. But the opposite is true. As soon as you close down one of the slits, denying the particles any choice, they somehow manage to reach places on the screen that they never reach when both slits are open.

What is the explanation for this bizarre behavior? No ordinary, commonsense picture of a tiny object explains the weird behavior a particle exhibits when it is given two opportunities to make a fact. It seems that the two opportunities—the two possibilities—somehow affect one another: They interfere with or bump into each other. But how can this be? No particle ever encounters the slits in the presence of another particle. Each passes through alone. Can quantum physics explain the apparent interference that this result suggests?

The answer is yes, but the answer changes our way of thinking. Suppose that instead of a single particle passing through the slits, what passes through is a wave. A wave doesn't behave like a particle. It would reach both slits at the same time and then break into two waves—one passing through each slit. This happens all the time when a real wave, such as an onshore ocean wave, comes to two openings, such as the spaces between three parallel supports for a pier. In this same manner, the two subatomic waves in the experiment's stream would travel separate paths to reach the screen, and they could interfere with each other when they came together again.

Waves are made up of moving, rolling hills and valleys. If a hill of one wave meets up with a valley of the other at the point where they hit the screen, the waves would cancel each other out. This would explain the places on the screen where no spots appear when both slits are open. Close down one of the slits, and the wave is not broken up into two parts that could interfere. *All* of the wave reaches the screen after passing through the single slit, so more points are recorded on more places on the screen.

The wave description turns out to solve the problem. Indeed, quantum physics was at first called *wave mechanics* because a wave description solved this and other dilemmas that occur whenever a subatomic object is faced with two or more possibilities. The possibilities always interfere with each other as if the particle was in some way a wave, suggesting that subatomic matter is really composed of waves. There were no particles, after all, in the stream. The stream is not made up of particles; it is made of waves.

But this isn't correct, either. When the waves arrive at the screen, they do not land everywhere on the screen, the way an ocean wave washes up on the beach at many places at the same time. Instead, each wave somehow "hits the beach" at a single spot. In other experiments, called particle-scattering experiments, involving subatomic and atomic particles traveling through space, the same thing turns out to be true. The final outcome is always that a particle leaves a track—a spot somewhere—yet travels through space as if it were a wave. Thus the waves are somehow particles, after all.

This behavior of subatomic matter when confronted with two or more possibilities is called the *wave-particle duality*. But giving it a name doesn't solve the problem it poses. We are still faced with a mystery—provided we believe in a strictly material world where particles are particles and waves are waves. The notion that the same entity can be both a particle and a wave opens up to us a new view of the world. To better grasp how this new quantum world behaves, it's helpful to have a brief summary of quantum-physics laws.

You may want to refer back to this short list if our later discussions become confusing.

1. Quantum physics deals with a new kind of object called a *possibility*-wave, and everything in the world, including your brain and your mind, has a *possibility*-wave.
2. *Possibility*-waves travel both forward and backward in time.
3. *Possibility*-waves can be superposed (added together) to create new possibility-waves.
4. *Possibility*-waves can be multiplied (squared) to create probabilities.

Parallel Realities or the Observer Effect?

In chapter 6, we examined the parallel universes theory, and I pointed out that an infinite number of parallel universes can exist side by side with no one usually the wiser. You might have wondered, If no one knows they exist, why posit them at all? It turns out that the presence of parallel universes not only resolves time-travel paradoxes but is integral in explaining the formation of the result of any observation. They can overlap and influence each other, thereby changing what we observe as a result. In other words, two alternative possibilities can interfere with each other. By "interfere" I mean that two, or more, possible outcomes somehow coalesce and produce a result that isn't present in either of the source universes taken separately, but appears in a new universe of its own, provided someone makes the attempt to observe it.

According to Deutsch, the two possibilities in the double-slit experiment, although describing only a single particle, are composed of two real, parallel-world particles—each particle really existing somehow in its own separate universe. Both universes are required to explain the interference. The pattern of hits on the record screen is not a simple compounding of particles passing through one slit or the other, but is the product of each particle interfering with its "ghost" particle in the other parallel world. Yet since the particles exist in separate and parallel universes, only one particle is ever found in any one universe. That would explain why only a single spot is observed (in each universe) after the particle passes through the slits. Thus in any single universe, even though the particle in the other universe is not present, the effect of its presence mysteriously changes the course of the observed particle's history and its final destination.

Parallel universes are not the easiest things to contemplate, and until recently many physicists preferred other interpretations of quantum physics. In the Bohr interpretation (named after Niels Bohr, a major contributor to the discovery of quantum physics), the observer of an event, such as a particle striking a screen, *causes* the event to occur. Bohr said that the simple act of observation changes quantum events, turning them from *possibility*-waves into probabilities. When an observation takes place, the object under scrutiny is thought to suddenly "pop" into existence. This has come to be known as the *observer effect*.

Today the interpretation of quantum physics still remains an open question, with as many as eight or more differing interpretations.[4] As I see, it there are essentially two schools of thought: Either the particles of matter do not exist when they are not observed and are only present when an observation takes place, or the particles do exist in an infinite number of parallel universes with a single observer branching out as parallel beings in all of them. In some sense these schools of thought may be saying the same thing, but this possibility is by no means clear. So far no experiment can tell the difference between any one interpretation and another. The sudden appearance of an object probably seems just as mysterious as the presence of parallel universes. It's most likely a measure of the limitation of the human mind, which thinks in concrete terms, that the quantum world appears paradoxical and mysterious to our thinking.

Let's revisit the double-slit experiment in light of the Bohr interpretation. Before any *observation* occurs, nothing resembling a particle exists. All we have are unobserved possibilities present as microscopic ghost waves. When an observation takes place, even before the observer has determined the actual value of the data he or she seeks, the particle suddenly *jumps* into reality. A single spot arrives on the screen and the *possibility*-wave becomes a probability.

Probabilities deal with real things. For example, after flipping a coin and then suddenly covering it with your hand, even though it has already landed, since you, the observer, have not seen the result yet, the coin still has a heads-up probability of 50 percent. Similarly, according the Bohr view, in the double-slit experiment, the experimenter may not know where the spot occurs, but she knows it has arrived. Before she examines the screen, she has in her mind the probability that the spot will be somewhere in the vicinity of the screen she explores. So the observer effect doesn't necessarily provide a determined answer for the observer, but it does acknowledge that the object in question is "out there," waiting to be discovered. Whatever the result will be, the event of its existence is not in question. Whereas before observation, the object in question can not be said to be "out there" at all. It resides in a mysterious world of *possibility*-waves.[5]

In the parallel universes interpretation, on the other hand, there is nothing special about the observer's point of view. When an observer observes an atomic object, the object changes by splitting into parallel worlds—but so does the observer. The parallel universes interpretation in fact explains the observer effect—the impact an observer has upon a physical system simply because he or she observed it: Nothing magical happens. The observer simply becomes part of the universe(s) in which the observation takes place.

In summary, the Bohr interpretation says: We don't know how an act of observation really takes place, but it can be imagined as the collapse of the wave to a single point. In the parallel universes interpretation, no collapse takes place. Instead, all possibilities arise in separate parallel universes. The *possibility*-wave is a means to take into account the interference potential of any universe with any other. Thus, when we say an object can move along alternative trajectories in parallel universes, we are saying the same thing as the object has a *possibility*-wave in one universe. Deutsch clearly had this in mind when he began thinking about quantum computers. Deutsch's interpretation says: It's possible to construct a quantum computer that operates without any collapse effects of observation. If one does not make any attempt to look into its operation, the quantum computer will perform as if it were in parallel universes rather than in a single universe. Ultimately, when an observer looks in for a result, one can then treat the effect of this observation as if it were a collapse. The observer simply enters into as many universes as there are possible outcomes from the quantum computer. Since he doesn't care about the other "ghost" quantum computers, the outcome to him appears as if a collapse had occurred. In effect, the collapse can be put in the hands of the user.[6] As I hinted above, this device uses the weirdness of quantum physics to carry out special calculations and predictions. Deutsch reasons that without using the parallel-universes interpretation, it would be difficult to fully understand how a quantum computer would operate. Hence, he has become a strong advocate for this interpretation of quantum physics.

Deutsch in a later paper even used the parallel-universes theory to show how a quantum computer's operation is completely consistent with the operation of a time machine.[7] That later paper is the basis for the discussion in chapter 6 of how the parallel universes theory resolves all time travel paradoxes.

As I mentioned, quantum computers are capable of calculating a new and mysteriously unobservable form of information. They work not only with numbers but also with number possibilities called *qubits*, meaning "quantum bits." To learn more about qubits, you'll want to read the appendix. There I explain in more detail how ordinary computers produce the amazing real number results that they do and how quantum computers do what they do. If you are curious, you may want to a look at the appendix now.[8]

Setting up the Quantum Computer for Time Travel

The remarkable thing about quantum computers is that they operate in parallel universes! Thus a quantum computer and its clones, each in its separate world, carry out parallel calculations. That's what makes them so powerful. You not only get to use the results of one computer's work, you get to use the results from an infinite number of them. To get our time machine working, we need to get the parallel-universes quantum computers working in harmony with each other, so that their outputs produce a special superposition of number possibilities. Let me call this special superposition "state-S." As we shall see shortly, there are two possible special superpositions of interest.

Since these number possibilities are never seen, you might wonder why we even believe they exist. As we saw earlier, we can't explain even the simplest atomic experiment, such as the double-slit experiment, without taking unseen parallel alternatives into account. When any observation is made, the observer becomes part of the experiment and, for him, the results appear consistent with the observer effect. Hence he doesn't see the presence of these unseen parallel universes, which come about through the superposition of possibilities.

In the double-slit experiment, if we had looked to see which slit the particle went through, it turns out we would not see the interference pattern on the second screen that we would have seen if we hadn't looked at the slits. The reason is that our observation would catapult us into two parallel universes wherein the particle would go through one slit or the other, with us following each trajectory. Only by not looking, not entering into the parallel universes of the particle as it makes its way to the second screen, do we get to see the interference effect when we just look at the second screen.

A quantum computer is no exception to the rule; its registers must behave just as the different pathways do in the double-slit experiment. If we don't look at the registers of the quantum computer, it will maintain state-S without any trouble. But if we do look at the register, we will see it jump to a particular number within the superposition. The quantum computer will no longer be in state-S. In the parallel-universes interpretation, we are split into as many number-possibility universes as make up the quantum computer's state-S. As far as we are concerned after we look in, we will not have access to any universe but one, and we will experience the observer effect.

So let's suppose we put the quantum computer into this state-S and remember to not look at the register. The next step is to consider the time machine itself and then what happens when we hook them together.

AHARONOV'S SPHERE OF MANY RADII

To envision how we can construct the time machine, we will look at a new idea chiefly developed by Israeli physicists Yakir Aharonov and Lev Vaidman, who have long been known for their rather ingenious ideas utilizing the unusual implications of quantum physics. In 1990, they and their team proposed a bizarre kind of time traveling device in which the traveler moves through time either backward or forward while the remainder of the universe remains on usual time.[9] In a review article written the following year, Vaidman explained in greater detail how the device would actually perform.[10] The idea is to put the time traveler inside of a giant, massive spherical hollowed-out shell and let him sit there for a period of time while the sphere is connected with a quantum computer that is in the special parallel universes state-S. Then the time traveler exits the device to experience the world while he is in a new state.

Spherical shells massive enough for this purpose are hard to come by. We need to imagine that one could be built, or that sometime in the future a large spherical cavity could be dug out inside the moon of a distant planet somewhere. Regardless of how it gets built, the key idea for this time machine comes from the general theory of relativity, which says that gravity slows time (see chapter 5).

Even if the sphere were not hooked up to a quantum computer in state-S, the time traveler, while inside the shell, would undergo a very slight gravitational time shift relative to the environment outside the sphere. This shift is due to the increased but constant gravitational potential field, made by the massive shell's presence, at all points inside the sphere.[11] It turns out that the potential field strength, and thereby the amount of shift, depends on the radius and the mass of the sphere—the smaller the radius and greater the mass, the greater the shift. On the other hand, increase the radius or decrease the mass, and the time shift decreases.

This gravity potential field yields what has been called a *classical* general relativistic time dilation (meaning that any clocks in the sphere slow down). For a spherical shell with the mass and radius of the earth, the time inside would slow down a mere 0.7 nanoseconds (less than a billionth of a second) each second. If the same amount of mass could be compacted into a spherical shell with a radius of, say, fifteen feet, that slowdown would increase to about a millisecond (thousandth of a second) each second.

Consequently, just by hanging out in the sphere the traveler would experience less time passing—a tiny, slowing down time shift relative to the time outside the sphere. Assuming that we could successfully build a thirty-foot-diameter massive spherical shell of the kind described above, our time traveler would only lose three seconds each fifty minutes. In just under two years, he would age one day less.

So small a time shift seems hardly worth the effort. But Aharonov and Vaidman found a surprisingly clever way to amplify the effect of the sphere's mass and radius, producing, in principle, a time shift as large as desired. They would connect the quantum computer in the special parallel universes superposition state-S to the sphere. As I described above, with regard to the observer looking into the double-slit experiment and thereby splitting into two slit universes, bringing the sphere into connection with the quantum computer causes the sphere to split and enter each parallel world that the quantum computer is in with a slightly different radius.

Each number possibility in the quantum computer connects up with a different possible radius for the sphere. That is, for each parallel-universes state in the computer making up the superposition state-S, the sphere's radius would take on a specific value. However, this in itself is not enough. Since the time shift varies with the radius of the sphere, inside the multi-radii sphere, the time traveler would experience different amounts of time passing simultaneously. In essence, in one parallel universe he would experience one time shift, while in the others he would experience different time shifts. There would be no overall effect on him because he would in effect be split into parallel universes with no overall consequence.

Aharonov and Vaidman, however, saw a way to get the universes to coalesce into a single universe by making the quantum computer enter a new *state* that I'll call "state-T," which, like state-S, is also not a number but a special superposition of number possibilities different from state-S. Consequently, all of the tiny possible time shifts are superposed together, producing one gigantic time shift. Furthermore, in contrast to the classical general relativistic time shift, which only slows down time, the quantum computer and sphere could produce either a positive or a negative time shift, resulting in the time traveler moving significantly forward or backward in time.

Whether he moves forward or backward depends on state-S. There are two states-S of interest. One I call "state-Sp," where "p" means "positive," and the other, "state-Sn," where "n" means "negative." When the quantum computer, in state-Sp, connects to the sphere, the sphere enters parallel worlds with a "positive" temporal disposition, while in state-Sn, with a "negative" disposition. This means the traveler can choose before entering the sphere which way he wishes to travel in time by setting up the computer in either state-Sp or state-Sn.

Here is where time shifting gets even weirder. Once we grasp what it means, we will be able to see how the mind plays a role in time traveling. The traveler's quantum *possibility*-wave—the means any person uses to determine specific knowledge about expected events—shifts to a time either earlier or later than the present time as observed by people external to the sphere. That is, the time shift has to do with shifting the time traveler's *possibility*-wave.

In chapter 8, I'll explain more fully how these *possibility*-waves behave. As I mentioned above, for now all we need to remember is that all things in the universe(s), including people, brains, and their minds, have *possibility*-waves. We only need to know that *possibility*-waves eventually provide probabilities—numbers that enable people to make predictions of the future. They describe not only the probability of certain events but also how the person thinks about and expects to experience those events. If the traveler were on his own in free space, as it were, watching a series of events take place, his *possibility*-wave would evolve naturally enough into the future, changing with time as new information pertaining to the event became available. For example, suppose he were expecting to experience a meeting with a friend or the outcome of a horserace. His *possibility*-wave would naturally change and evolve in time, enabling him to predict these events with greater and greater accuracy as the future events unfolded before his eyes. However, since the time traveler is inside the sphere, his *possibility*-wave evolution will not be natural and can be severely distorted by a large enough time shift.

The chance for a successful operation of this machine depends critically on how often you can induce the quantum computer to transition from state-S into this special overlap state-T, in which the small time shifts add together to produce a giant time shift. Aharonov and Vaidman called this state-T of the quantum computer a *post-selection measurement* and explained that its occurrence is so rare that it would hardly ever happen. Putting the computer into state-T, after having it in either state-Sp or state-Sn, turns out to be a difficult task, because this particular superposition is very different from either state-S. This means that once a state-S is produced, the probability of state-T occurring is only very small, although not impossible. In fact, it is so unlikely that we might ask, why build the machine at all? If we could have known the time traveler's state of mind to begin with, we would have found that the small odds of his spontaneously shifting to state-T would be the same without the fancy equipment. Hence, it appears that nothing would really be gained by using a quantum computer and the sphere in the first place.

So why go to all of the bother? The answer has to do with the time traveler's state of knowledge. If we decide to shift the traveler's *possibility*-wave in time without the quantum computer or the massive sphere, we would need to know exactly what state his mind was in to begin with. We would need to know how he was planning to see the world, including himself in it. In other words, we would need to know specifically the traveler's *possibility*-wave. But human beings are very complex, so to determine this would seem hopeless.

That's just the point the designers realized. Their machine doesn't care what the traveler's *possibility*-wave is. It makes no difference what state it is in, for the sphere can shift the time traveler's *possibility*-wave without knowing what the *possibility*-wave was to begin with. This applies to any *possibility*-wave confined within the sphere. As long as the quantum computer is induced to enter the post-selection state, the traveler's *possibility*-wave will also shift.

To get a hold on what this all means, we need to suppose that the traveler enters the machine at some time and exits at a later time. Suppose that the traveler put the computer in state-S before he entered and got it to switch to state-T just before he exited. Suppose that he started with state-Sp desiring a positive time shift, and then after exiting turned the device off. Where would he be in time? The time traveler would find himself shifted rapidly forward to a later time in his natural evolution, while the external world ticked on in normal time. The time traveler's *possibility*-wave would now describe the traveler's state as if he was actually in the future. Certainly he would be out in the world of normally elapsed time, but he would have a precognitive ability allowing him to pick up on what will happen to him in the future. Thus the traveler would find his probability of success for predicting the future greater than if he had never entered the sphere. In other words, he would be experiencing a flash forward in time. He would see the future before it happened.

But there is another consequence of this flash forward: He would also age faster. By seeing the future he also goes to that future, as it were. He becomes older than his age as measured on clocks outside the sphere. If, for example, his flash forward took him ten years into the future, he would know about that future, but he would pay the price of being ten years older to do so. His friends outside the sphere would remark on how much older he appears, for example.

When it comes to backward movement through time, things get even stranger. Now we picture the traveler entering the device with the quantum computer in state-Sn, staying in it for a period of time, successfully switching the quantum computer to state-T, then turning the machine off and exiting it. When he enters the world again, it has naturally evolved in time and gotten older, accordingly. But not the time traveler. This time, depending on the magnitude of the shift, his *possibility*-wave shifts him backward to an earlier time in his natural evolution, even to a time before he came in contact with the device. However, he returns to a past state he would have been in, had he remained in isolation from the world and free of the sphere. Even though the external world ticks on in normal time, the traveler after switching the quantum computer to state-T would relive an in-isolation-from-the-rest-of-the-world past. By isolation from the world, I mean being cut off from outside stimuli. Although I didn't mention it above, when he enters the device he must also become isolated from the rest of the world until he exits, in order to make the device work properly regardless of a positive or negative time shift.

Time travel to the past using this device is a little trickier because of the isolation factor. Becoming younger would seem to be desirable, but let's think a little more about it. First, the traveler would get younger, but he may not necessarily return to his actual youthful self! Instead, he would return to an isolated past state of himself that would have also evolved to the state he was in before he entered the time machine, had he been in isolation in that past. His actual past state may have been quite different, and assuming he wasn't in isolation, most assuredly *was* different.

The machine actually shifts the *possibility*-wave of the time traveler to the past he would have had, if he had been isolated from the world-at-large. This is called the *counterfactual* past.

Let me give you an example. Consider the following proposition: Suppose on Monday, you met someone who through his remarks or attitude made you angry. After a couple of days pass, you feel calm and, on Wednesday, decide to enter the time machine. The machine now successfully takes you back in time to the counterfactual state you were in on Monday. But what does that mean? It turns out you won't go back to that anger-provoking state you were in when you met with that person. Instead, you will return to that state you *would* have been in, had you not met that person at all. Assuming you are the kind of person who normally doesn't get angry when you are isolated from other people, you would thus return to a relatively calmer period. In this way, the device could be used to wipe out bad memories.

Or, suppose you had in the past taken in a toxic substance or heard some words that angered you and had a toxic effect on your mind-body, making you ill. Could time reversal in the machine remove this illness? The answer is yes, provided the counterfactual past was not toxic. If you were well in the counterfactual past, then you would find yourself returning to a healthier state. Hence, the device could also be used to cure illnesses.

The PHYSICS *of* "EXTRAORDINARY" TIME TRAVEL

Do not dwell in the past, do not dream of the future,
concentrate the mind on the present moment.

—Buddha

A quantum mechanic does not fix your car,
but he may fix your time machine.

—motto of the Physics Conscious Research Group

The discussion in the last three chapters concerning parallel universes and contraptions for time travel, including traversable wormholes, quantum computers, and finally Aharonov and Vaidman's sphere, was meant to lay the groundwork for the discussion in the present chapter. Here we will see how possibility, probability, and time are related. And we will discover how *possibility*-waves provide a subreality-to-reality connection and how a simple mathematical operation called *squaring* changes these subreality waves into reality probability-curves—which are the basis of the cause-and-effect world that appears to us as reality. Though the concepts in this chapter may sometimes sound like science fiction or a metaphysical turn around the bend,

I assure you they are based on some very hard thinking by many physicists.

The key lies in understanding what actually gets shifted inside Aharonov and Vaidman's sphere of many radii: It is the *possibility*-wave of the time traveler that shifts in time. We will not worry about a device that enables a time shift to take place. Rather, we will look into what the time shift accomplishes in the mind. Once we grasp that, it is just a short journey to a mind yoga for time travel—a means to actually cheat time.

One insight to be gained is that time and possibility are intimately connected, far more intimately than could have been even thought about before the discoveries of quantum physics. The connection between time and possibility is very subtle and has to do with how possibilities change into probabilities when awareness enters into the picture.

Awareness implies something deeper than might be obvious. It implies an exchange in information resulting in a possible gain or loss of knowledge. Briefly, when you become aware of certain information, there is a gain of knowledge, and we know that as a consequence probabilities change. For example, when you observe that a flipped coin lands "heads-up," you know the probability for observing the coin "tails-up" becomes zero. Before you saw the coin land, the probability would have been 50 percent. So awareness certainly changes knowledge and thereby probabilities. Although it may be surprising, a change in probability can also result in a loss of knowledge, as I will explain below.

It may be useful to look at what I mean by *knowledge* and *awareness* in terms suggested by quantum-physics principles. Awareness is a two-pronged physical action—one, when a possibility changes into a probability and two, when a probability changes. A probability can change into a certainty or not; if not, it can become less certain or it can become more certain. I will at times use the word *consciousness* to mean the same thing as *awareness*. *Knowledge* is the outcome of acts of awareness. It can be thought of objectively as something physical. We experience knowledge as memory. That is, once we know something it enters memory. However, knowledge can change; it can be gained and even be lost, and awareness can also change knowledge by its action of changing probabilities and possibilities.

Awareness or consciousness can change a more certain probability into a less certain probability. In such a case, an observer may actually lose knowledge of one kind but gain knowledge of another. When this occurs, an observer has changed knowledge. Knowledge thus represents a specific outcome of such an action and is subject to change. We may know something about an object one minute and, in an attempt to gain more knowledge about the object, we can actually lose some of the knowledge we originally had. This change in knowledge occurs in quantum physics because of choices an observer has. These choices occur in complementary pairs so that one choice always precludes the other. Thus, if choice A is made and an attempt is carried out to observe an object, any information or knowledge about the object gained by choice B will be lost. In quantum physics, this is known as the *complementarity principle,* and because of it we must always deal with possibilities and probabilities and how these change.

A change in the physical situation can also change possibilities. In what follows, I hope to make clear how changes in the physical environment and changes in awareness differ in the manner in which they change possibilities. First, we are asking about *possibility*-waves and what role choice and complementarity plays in changing them. Let's again look at the double-slit experiment of chapter 7 to see how a change in the physical situation changes possibilities and probabilities and how such changes change knowledge. When we open or close down a slit, we change the physical situation and the *possibility*-wave representing that situation. There, by choice, we changed the *possibility*-wave by changing the number of open slits in the screen. With observation of a single slit open, we gained knowledge of the pathway taken by the particles as they made their way to the recording screen. When we opened both slits, we lost this knowledge; we no longer knew by which pathway the particles made their way to the recording screen. In this example, observing the record made by the particles with one slit open is complementary to observing them with both slits open. In fact, this is nothing more than the wave-particle duality discussed in many popular texts dealing with quantum physics, wherein one can choose to observe either the wave or the particle aspects of objects by changing the way in which the experiment is carried out.

Certainly changing the physical situation as indicated above can change a *possibility*-wave, for, after all, the *possibility*-wave represents the physical situation. In the remainder of this chapter, we will explore another way in which such a change can occur. We will see how a *possibility*-wave can change as a result of making different choices in the ways we go about observation. I call this simply changing awareness. Simple, yes, but such a change is profound in its implication.

If we can learn to alter our awareness in a certain way—that is, learn to become mind yogis, we can change possibilities, alter outcomes, and even enable outcomes to "flow" as new *possibility*-waves that eventually emerge as new physical events.

IS THE *POSSIBILITY*-WAVE REAL?

To begin with, let me review for you what a *possibility*-wave represents by using an example. Ordinary computers use *bits*—binary numbers that take on the value of 0 or 1. In chapter 7 we saw that quantum computers work with qubits. Qubits (which, you recall, are quantum bits) are pure number-*possibilities* that take on the value of either 0 or 1 whenever they are observed, but when unobserved have no actual value at all. A qubit can be imagined as an arrow that is pointing not up (indicating the value 0), or to the right (indicating the value 1), but to any position in between. While a bit may be suddenly flipped between 0 and 1, an unobserved qubit changes its position continuously and, following the laws of quantum physics, rotates.

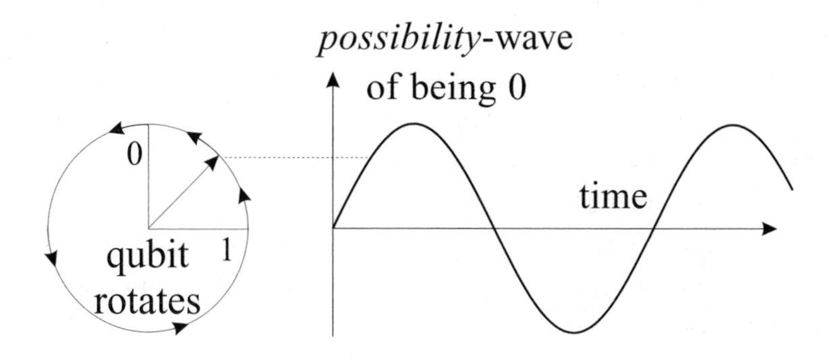

Figure 8.1. *An unobserved qubit continuously rotates in time and traces out a* possibility-*wave.*

Figure 8.1 shows how, as time passes, the qubit rotates in a complete circle. As it proceeds around the circle, the arrow's tip also traces out a curve in time shown in the figure to the right of the circle. Hence the possibility of the qubit having a value of, say, *0* depends on *when* you look at it. As you see, the *possibility*-wave oscillates between a positive maximum value and negative minimum value.

In the figure, the qubit's *possibility*-wave represents the chance of the qubit having the value *0* at any instant. The laws of quantum physics can only determine, in this way, the probability of the qubit having this value. One cannot predict for certain what the value is; all one can predict is what the value is likely to be. Since there are two possible values, 0 and 1, the qubit has only two possible ways to go.

The qubit's *possibility*-wave oscillates between positive and negative values. But what does it mean that the *possibility*-wave has a negative value at times? Two ideas come to mind. It could mean that the value of the qubit is negative at that moment. But we would never see the qubit with a value of *-0* or *-1* or any negative value at all. Or it could mean the probability is less than 0 in that moment. But that is even more mysterious. How could a probability be negative?[1]

This is where quantum physics and common sense part company and physicists find much to disagree about. Some physicists believe that a *possibility*-wave is not real, in spite of its having a real effect in the world. They believe that the *possibility*-wave exists only in the mind, so that physicists can eventually use them to make calculations of probabilities. Perhaps the *possibility*-wave's negative number helps with the bookkeeping in some manner to balance the odds. For what else could be meant by a negative number *possibility*-wave?[2]

On the other hand, many physicists take the opposite tack and believe that *possibility*-waves are very real, even though they are invisible, sometime negative, and undetectable. I'll tell you about one of these physicists and his picture of *possibility*-waves later on.

Let's go into this negative number *possibility*-wave a little farther and how it can result in a positive probability. Mathematical science in general deals with probabilities—numbers that lie between 0 and 1—and provides answers to questions such as, how likely is the occurrence of an event. For a flipped coin, the event might be landing heads-up. If the coin is a "fair" one, then that likelihood or probability is obviously .50 (fifty percent). If the coin is weighted, that probability could be .75 or actually any number between 0 and 1, but never a negative number. Hence, how can the *possibility*-wave relate to a probability? To deal with this relation, physicists noticed that the *possibility*-wave when multiplied by itself—the square of the *possibility*-wave—always gives a positive number.[3] That led them to believe the *possibility*-wave does have something to do with probabilities after all.

In figure 8.2, we see a comparison between a *possibility*-wave and its square, resulting in the probability of seeing the qubit with value *zero.* As you can see, at no time is the probability-curve ever negative. But the *possibility*-wave still has a negative part. So again we ask, what could the meaning be of a negative *possibility*-wave? (The answer again takes us into a realm where common sense may elude us.)

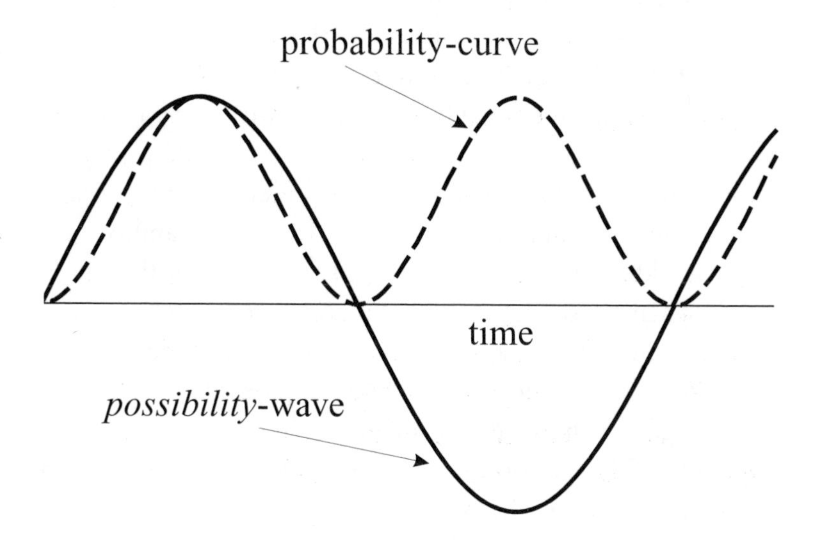

Figure 8.2. *A* possibility-*wave and its probability-curve.*

Let's consider another squaring example. It illustrates how the negative values of *possibility*-waves in a superposition of two or more waves change the probability of a physical event occurring. In chapter 7 we saw that *superposition* is when two or more *possibility*-waves add together. This situation arises whenever an object has two or more pathways to its final outcome. Suppose, for example, we look at our qubit again and imagine that, like in the double-slit experiment where a particle can travel by two pathways at the same time, the qubit can rotate in two opposite directions and that both rotational possibilities exist simultaneously, as shown in figure 8.3.

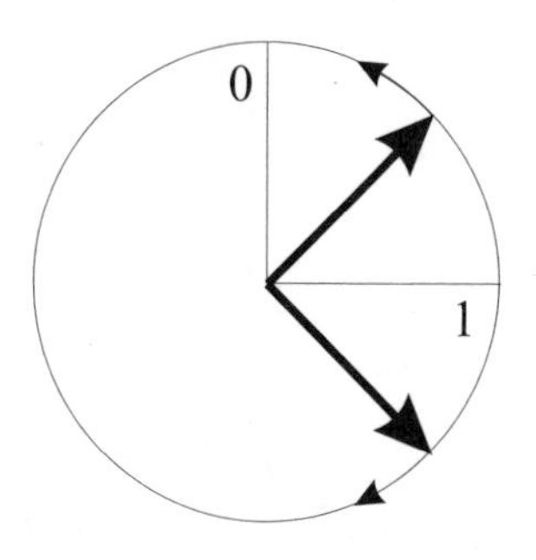

Figure 8.3. *Two qubit possibilities rotating in opposite directions.*

In figure 8.4 we see that the qubit's two rotational *possibility*-waves trace out competing curves, so that when you add them together they cancel each other out. That is, the positive part of one curve always comes together with a negative part of the other curve and vice versa. To keep the bit value and the probability values separate, I'll use the word *zero* to refer to the bit value of the qubit and the number 0 to refer to the probability value. When we add the two *possibility*-waves, we find that they add to 0. Thus, when we square 0, we get 0 and therefore a probability of 0 at any time. This means that there is no chance that the qubit will ever take on the value of zero. Since a qubit can only have one of two possible values, zero or one, and since we conclude that the qubit would never produce the bit-value of zero, it must produce the value one at all times.

But, suppose we squared the *possibility*-waves for each rotational possibility *before* we added them. Would the result still be the same? The answer is no. In this case, each *possibility*-wave would be squared first to yield exactly identical probability-curves, as shown in figure 8.5. When you add those two probability-curves together, they don't cancel each other out. Thus, it seems to be very important to know when to add probability-curves and when to add *possibility*-waves, for the results differ tremendously.

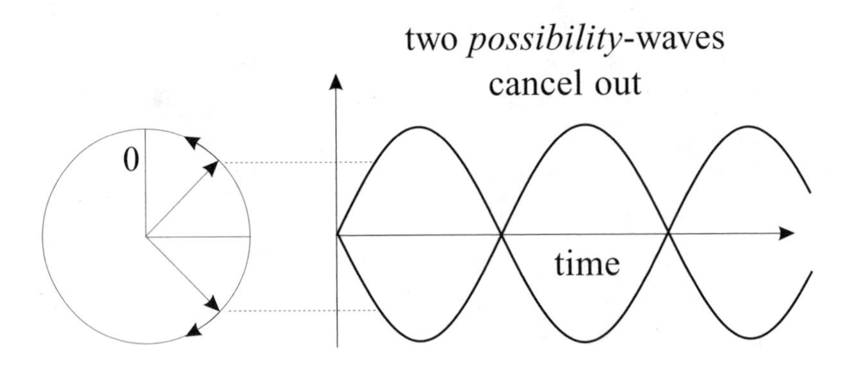

Figure 8.4. *Two* possibility-*waves can add and cancel each other out.*

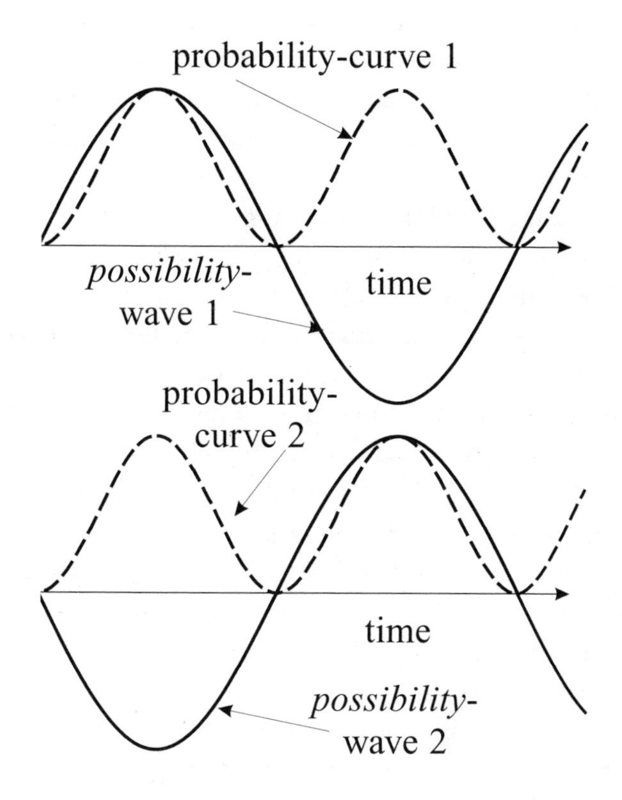

Figure 8.5. *Different* possibility-*waves can produce the same probabilities.*

IMAGINAL *POSSIBILITY*-WAVE AND REAL PROBABLITY-CURVES

As we see in figure 8.4, when a qubit's *possibility*-waves add together, they can produce different probabilities of the qubits having values zero or one. These *possibility*-waves are crucial to the operation of quantum computers. In fact, that is the whole point of designing quantum computers in the first place: to take advantage of this new kind of computer addition called the *law of superposition* in quantum physics (see chapter 7 and the appendix). When two opposite qubit *possibility*-waves are present, the probability for observing a qubit value is different from the situation when only one such wave is present. The two possibilities create an interesting situation, in that they can act together and affect quite strongly what takes place in the real world.

It is strange to think that adding two *possibility*-waves can produce a result that changes our perceived reality. It certainly seems unreal that a single qubit, or any object for that matter, can move in opposite directions at the same time. But according to quantum physics, all unobserved objects must behave in this strange way; they must simultaneously move in as many directions as is possible for them. Because the unobserved qubit can move in both directions at the same time, it will do so. When we see that adding two *possibility*-waves together changes both our observation of reality and reality itself, we need to consider that the two *possibility*-waves might also be "real," in spite of the imaginal quality of such ideas as negative *possibility*-waves and the fact that we never see them.

By comparison, if we are dealing with two or more probability-curves, nothing very mysterious results. We know how to think about this situation; we simply add up the curves. Each curve represents a probable reality—a way in which a result can occur. If we attempt to see how likely it is that a certain result will follow, we add up all of the ways it could happen. For example, to arrive at the probability for throwing the number 7 with a pair of dice, we add up the probabilities for all of the different ways this could result. Since we have two dice and each die has six faces, there are 6x6, or 36 possible ways for the pair of dice to show any number. In this example, we will take for granted that the probability-curve for each number does not vary through time. We then add up all the ways that the number 7 could be produced. We could get 1+6, 2+5, 3+4, 4+3, 5+2, or 6+1, where the first number corresponds to die-1 and the second to die-2. Altogether, there are six different ways, each way having a probability of 1/36. Hence the probability of getting a 7 comes out to be 6/36 or .166667 (see figure 8.6).

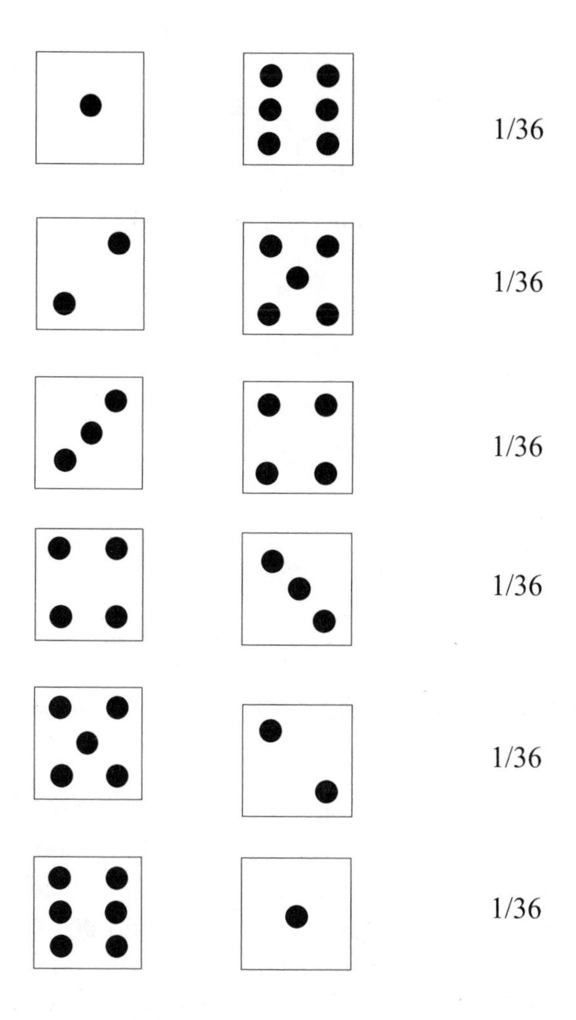

Figure 8.6. *Probabilities added to get lucky seven.*

Probabilities depend on real things—things we can count as "out there," not just in our minds. Hence when we deal with probability-curves, we need to only consider real outcomes and all of the "real" ways any desired outcome can occur.

Consider another example, a roulette wheel. As the wheel spins, the ball travels around the wheel until it falls into one of the wheel's notches. Here we'll imagine what the probability-curve looks like for the ball falling into a red slot.

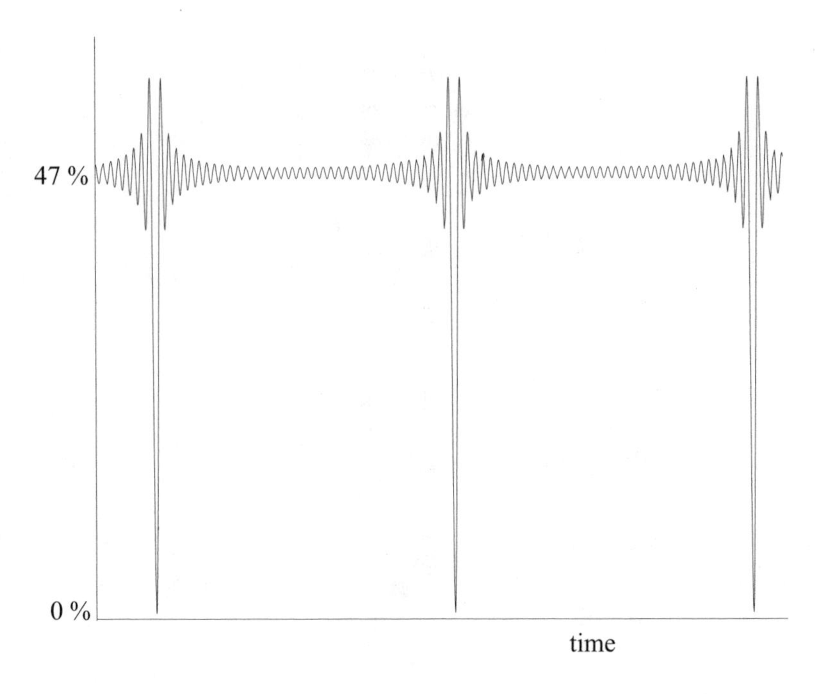

Figure 8.7. *A fictional probability-curve for "red" on a roulette wheel.*

If we want to determine the probability of the ball reaching a red number, we add up the separate probability-curves for each red number. Since there are 18 red numbers, 18 black numbers, and 2 green numbers, consisting of the 0 and 00—38 numbers altogether—each has a probability-curve peak no higher than 1/38 and a valley at least as low as 0 (assuming the wheel is fair). By adding up the probability-curves for each red number, we arrive at a probability-curve for red, which is 18/38 or about a 47 percent chance for a red outcome (see figure 8.7).

For a fair wheel, the probability-curves for all of the 18 red numbers when added together produce a fairly flat curve across time. As the wheel turns, the probability for a roulette ball falling into any red number may indeed vary slightly in time. The curve has small ripples, but, at nearly all times, it stays around the value of 47 percent. This indicates that the probability for getting a red number doesn't vary much from this value.

But to make things more interesting, let's suppose the probability for producing a red number in roulette can change dramatically in time, as also shown in figure 8.7. The curve rushes down to zero at periodic moments, indicating that at those times there is no chance of the ball landing on a red number. How could this actually happen? Perhaps there are tiny magnets in the red notches that the operator turns on from time to time, causing the roulette ball to be pushed out if it starts to fall into a red notch. This would tend to favor black or green. Since the house's cut often comes from green numbers, perhaps the house has magnets installed in both the red and black numbers, in which case the probability-curve for black would appear as shown in figure 8.7 as well.

But even with the house cheating, in no way do the different probability-curves for the red numbers when added together create a canceling effect. Probability-curves *never* cancel each other out, because they are never negative. On the other hand, *possibility*-waves are capable of canceling each other out and nearly always do so, as we saw in figure 8.4.

There we saw just two *possibility*-waves adding together and canceling each other out. Now consider what happens when we take all 18 *possibility*-waves, each wave representing a different red number, and add these together to produce one *possibility*-wave—the sum of them all. Suppose a kind of giant house conspiracy exists and that the *possibility*-wave for each red number has been put into such a superposition. Since these waves have both positive and negative parts, we can understand why they cancel each other out and why the curve in figure 8.8 looks as it does. It shows the result obtained from adding up all 18 "red-number" *possibility*-waves to get one "red" *possibility*-wave. If the house could control the situation as shown in the figure, there would be little chance for red to show up at all except for certain periodic times.

How could the house produce this result? Perhaps the wheel is governed by quantum physical laws such that the red numbers on the wheel act like slits in the double-slit experiment. Just as we

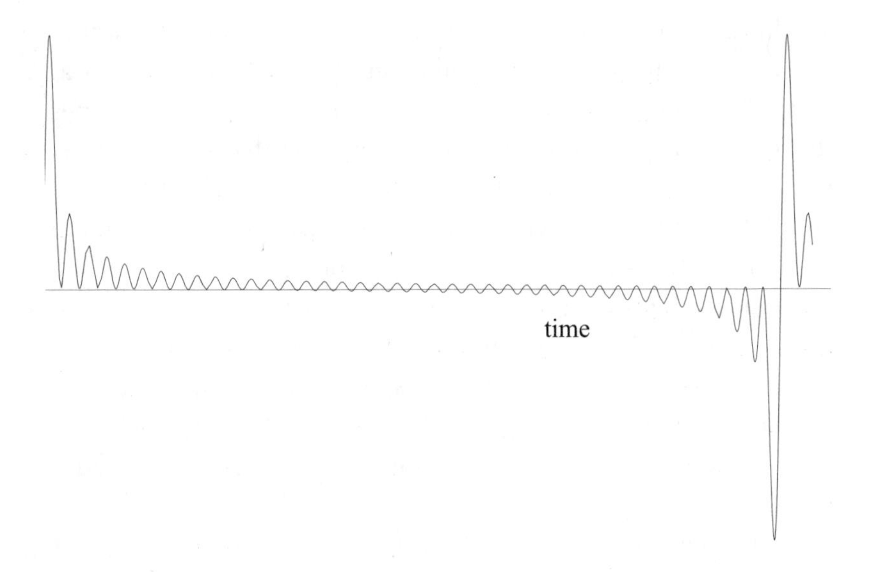

Figure 8.8. *A* possibility-*wave for "red" on a roulette wheel.*

saw in chapter 7, figure 7.3, when two slits are available for a particle to pass through, the two *possibility*-waves interfere with each other, resulting in strips on the collection screen where the particles do not go. This means the probability-curve for the particles has zero value in these areas.

With the 18 red numbers on the roulette wheel, instead of 2 slits, a greater number of canceling effects result, giving rise to figure 8.8. The probability-curve obtained by simply squaring the *possibility*-wave shown in figure 8.8 for observing a red number turns out to be nearly zero for most of the time the wheel spins.

Figure 8.9 shows the resulting probability-curve for red reaching near certainty at periodic times and virtually zero for most of the time. As we see, at periodic times the house is tilting the odds in its favor by allowing interference between the various red number *possibility*-waves, resulting in a peak, only at certain moments. If you are lucky enough to place your bet when the probability-curve hits a maximum, you will certainly win.

The important idea here is that probability-curves can appear quite different from the *possibility*-waves that make them up. The question then naturally arises: When do we add *possibility*-waves and when do we add probability-curves?

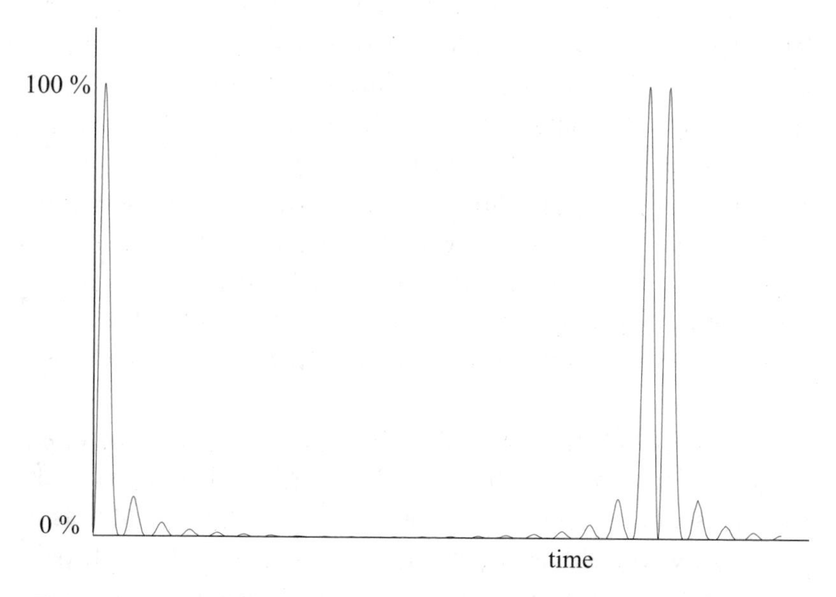

Figure 8.9. *A probability-curve for "red" on a roulette wheel obtained from adding all "red"* possibility-*waves and then squaring.*

ADDING *POSSIBILITY*-WAVES AND ADDING PROBABLITY-CURVES

A wave of possibilities sounds like something quite magical. Since it is a wave, we assume it does all the things that waves do—it undulates, vibrates, and moves through space and time. But does a *possibility*-wave really move through space and time, or does its motion exist solely in our imagination?

It certainly appears that situations where objects can follow multiple paths to a single outcome cannot be real. An object must follow a single path from here to there; the notion of multiple possible paths is only a useful mental construct, and in our imagination, anything is possible. Thus, when a *possibility*-wave encounters a situation that is *logically impossible in the real world,* we can still use our imagination to envision how it might behave. For instance, remember the metaphor in the previous chapter: When an ocean wave encounters the spaces between three parallel supports of a pier, it splits apart and enters each of these spaces. So we envision that the imaginal *possibility*-wave also splits and goes running off after each possibility simultaneously.

But in the "real world," we don't see *possibility*-waves at all. We don't see an object splitting apart each time two or more possibilities arise—for example, when flipping a coin, rolling dice, or spinning a roulette wheel. For that matter, we really don't see the possibilities splitting in the quantum physics double-slit experiment or even when we consider qubits. We *always* see only a single outcome.

When we tally up the results of our observations of anything, tiny or not, what we see are "real" things. We deal with probability-curves, not *possibility*-waves, and we determine the odds as if *possibility*-waves did not exist. Yet they apparently do exist. Not only that, from the viewpoint of quantum physics, they are essential to every perception we make. In the end, it is the squaring of all possibility-waves that allows us to make sense of the world. I mean *sense* in two ways: the ability to directly perceive that world through our common senses, and the ability to order and understand what we perceive.

Let's review what we know about *possibility*-waves and probability curves. We know that probability-curves add up to produce outcome probabilities. We know that when *possibility*-waves add up, they either cancel or reinforce each other, each option producing a very different outcome probability. When we consider *possibility*-waves, we add them and then square the result to get a probability-curve. But when we only consider separate probability-curves, we add them to get a single probability-curve. Consider the following questions:

1. What determines whether we add probability-curves after squaring *possibility*-waves or add *possibility*-waves before squaring them to yield a probability-curve?

2. In other words, when do we add probability-curves and when do we add *possibility*-waves?

3. Put slightly differently, which comes first: add and then square, or square and then add?

Since *possibility*-waves are apparently a figment of our mind, it seems the answer to the first question can be stated simply enough: our mind. And since probability-curves do correspond to reality—that stuff "out there"—the answer to the second question must be: We add probability-curves when we become conscious of "out there," so it should logically follow that we would add *possibility*-waves when we remain conscious of "in here" rather than "out there." The answer to the third question is: When we deal with the world as we imagine it to be, we add and then square. But when we deal with the world outside of our minds, we square and then add.

These answers may seem reasonable enough, yet there is a subtle kind of trap here. The trap is concerned with just what we think the mind *does*. We are used to thinking of the mind as holding sway in its own internal world, separate and distinct from the objective world "out there"; the mind can't affect what happens in the objective world. We assume, as Aristotle pointed out long ago, that mind and matter are different categories of things and that one cannot have direct action on the other. Most philosophers today consider this two-category analogy to be true—the mind cannot directly influence matter, nor vice versa. Certainly when we "make up our minds" to do something, we know what it means to put that idea into action by moving objects, including ourselves, about in the world. In that sense we're familiar with the mind influencing the world indirectly. But what would it mean for mind to affect something *directly* in the world? Is that even possible?

Quantum physics has offered us a conundrum, for it seems to be telling us that mind and matter are directly connected through the hookup between the *possibility*-wave and the probability-curve.

Certainly we see evidence of this connection in this business of adding *possibility*-waves and then squaring them to get probability-curves, or of squaring the *possibility*-waves and then adding the resultant probability-curves. I would like to add mind into the equation, so to speak, and simply say that *this* is what mind *does:* It converts *possibility*-waves to probability-curves by performing this squaring operation, which then produces probabilistic effects in the real world. Mind you, this is only a speculation, although certainly a reasonable one when we look at what quantum physics has to say about reality.

CAN *POSSIBILITY*-WAVES GO BACKWARDS IN TIME?

Are *possibility*-waves real? That is the fundamental question that concerns us here. I have suggested that in some sense they are not real; yet they do have very real consequences. However, many physicists believe that *possibility*-waves are real and exist in some ghostly form or manner affecting the outcomes of experiments. A main proponent for this point of view is physicist John G. Cramer, who proposes that *possibility*-waves really travel through space and time in both directions![4]

The normally accepted point of view, originating with Bohr and in which some physicists believe, imagines that when any observation occurs the *possibility*-wave paranormally squares itself, producing a probability-curve. In explaining this squaring operation, this school of thought usually evokes some form of "magical wand" to carry out the squaring operation, yet no one can find a quantum rule spelling out how some sort of physical agent could ever appear.

Recognizing this limitation, Cramer asked: How does this squaring occur? He noticed that this operation is a little different from just multiplying the wave by itself. To compute the probability of the event, the wave must actually be multiplied by another wave that is nevertheless nearly the same in form and content as the original wave. This other *possibility*-wave, for mathematical reasons, is called the *complex-conjugate* wave, and it differs in a subtle way from the original *possibility*-wave.[5]

Multiplying two mathematical entities together to obtain a single number is quite common in physics and, for that matter, in your daily life. For example, to determine the distance you travel, you multiply the speed at which you move by the amount of time the journey takes. Or to determine the cost of apples at the grocery store, you multiply the number of pounds of apples you buy by the price per pound. You use common sense and the accepted laws of physics or commerce.

However, even though quantum physics is quite rigorous, nowhere in it is there any law explaining what occurs physically when a quantum wave is multiplied by its complex-conjugate. Nowhere is the complex-conjugate wave given any physical significance, except for a funny little quirk: The complex-conjugate wave happens to be a solution to the same equations of quantum physics solved by the original *possibility*-wave, provided that in writing those equations you let time run backward instead of forward![6]

Now, the *possibility*-wave has never been seen, although Cramer and other physicists may have a great deal of faith in its existence. It is just a solution to an equation. But if it is a real physical wave—one that exists and propagates through space and in time—then the conjugate wave, which also has never been seen, is not a mystery, provided you are willing to borrow an idea from science fiction and let it run backwards through time. So goes the argument: If the quantum wave is a real wave, then the conjugate wave is also a real physical wave,[7] but with a twist in time.

Any wave, including a *possibility*-wave, must move from one place to another, and it must take some time to do so. We can imagine the wave propagating through space much as a ripple moves across the still surface of a pond after a stone has been dropped in the water. We picture it expanding ever outward.

How, then, would a time-reversed wave look? Here we need to be careful in our thinking. To really picture a time-reversed wave, you need to envision going backward both in space and in time. Using the example of the pond again, a time-reversed wave would suddenly appear at the pond's boundary and would squeeze in on itself, ever contracting until it collapsed to a single place where the stone hit the still water. We can simulate this observation by watching a movie of a wave run backwards through the projector.

Thus the conjugate *possibility*-wave travels in the opposite spatial direction as it goes back through time, eventually reaching the original *possibility*-wave's origin. We imagine that at every point along its way it meets up with the original wave coming forward in time. The two then combine in space. In physics, the conjugate wave is said to "modulate" the original wave. Wave modulation is quite familiar to scientists and engineers working in radar, radio, and television. When you tune your receiver, television set, or radio to a station, you are picking out of the air a certain well-defined and quite narrow band of transmission frequencies sent by the broadcasting station. The central part of this band is called the *carrier frequency*. However, that carrier frequency is not what you hear or see. Usually the carrier wave frequency is much, much higher than the frequencies for audio and video signals. The information making up the sounds you hear and the pictures you watch are carried piggyback by the carrier wave. That means the information and carrier waves are multiplied together (see figure 8.10). The information you see and hear is simply wave forms that modulate or cause the strength or the frequency of the carrier wave to change because of the multiplication.

When the conjugate wave modulates the original wave, mathematically this is nothing more than the product of the two waves multiplied together. Since the waves are identical in form, this multiplication is, in effect, squaring. In order for any event to occur, both quantum waves must be simultaneously present, one modulating the other. As Cramer explains it, when the future-generated conjugate wave propagates back through time to the

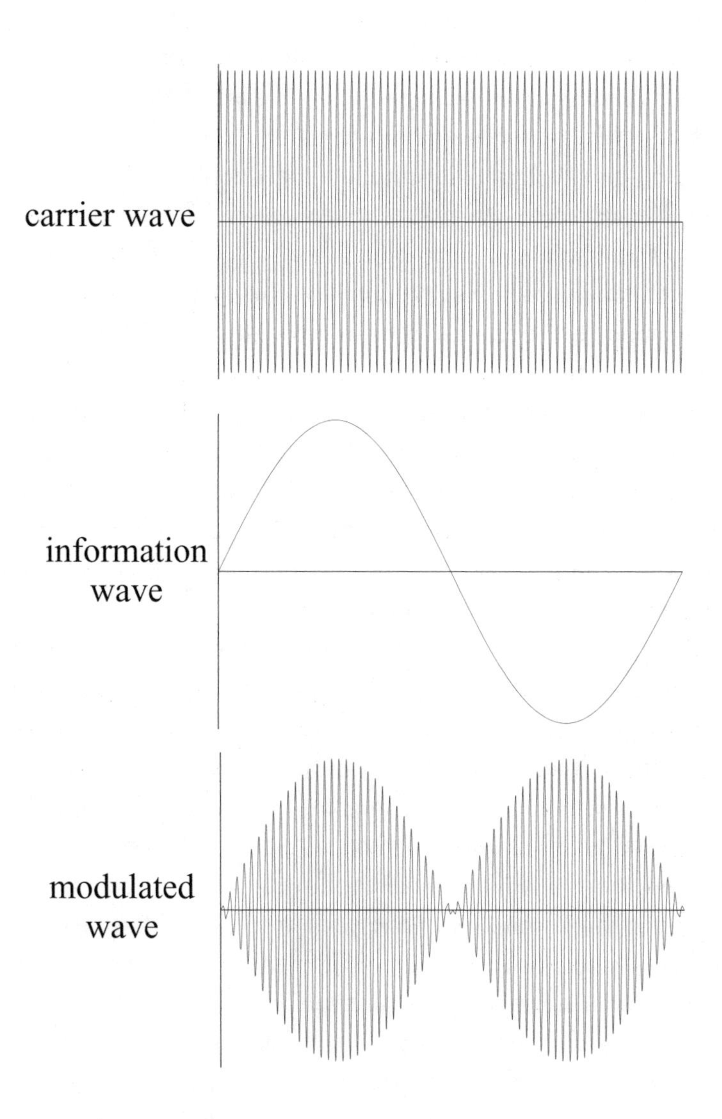

Figure 8.10. *Wave modulation.*

origin of the quantum wave itself, it meets the original quantum wave. Then in space and time the two waves multiply, and the result is the creation of the probability-curve for the event occurring in space and time.

Cramer calls the original wave an "offer" wave, the conjugate wave an "echo" wave, and the multiplication of the two a "transaction." A transaction occurs—involving an offer and an echo—much like that between a computer and a peripheral device, say, a printer or another computer over a telephone line. In these examples, an offer wave is sent to a receiver. The receiver accepts the offer and sends confirmation back along the same line.

In the *possibility*-wave/complex-conjugate-wave sequence, the exchange is the same except that because of the time-reversal, the offer and the echo cyclically repeat until the net exchange of energy—and other physical quantities that will manifest—satisfy certain "reality" requirements. These include the conservation laws of physics and any other restrictions imposed on the quantum wave, which are known as boundary conditions. When all criteria are met, the transaction is complete and the *possibility*-waves are changed into probability-curves.

If we take Cramer's interpretation seriously, we have a whole new picture of time with regard to quantum events. Every observation is both the start of a wave propagating toward the future in search of a receiver-event and itself the receiver of a wave that propagated towards it from some past observation-event. In other words, every observation—every act of conscious awareness—sends out both a wave toward the future and a wave toward the past. Both the beginning of the wave and the end appear in our minds—our future mind, our present mind, and our past mind. Two events in normal, or serial, time are then said to be significantly connected, that is, meaningfully associated, with respect to each other, provided that the transaction between them conserves the necessary physical constants and satisfies the necessary boundary conditions.

But an interesting problem remains: Which future event sends back the echo wave? Cramer believes that only one future does this—the one producing the echo that happens to have the best chance of forming a successful transaction with the present. But what about all the other possible futures? How could future events with less-than-best probabilities ever occur?

Some Reflections from the Future and the Past

Here I would like to present a new idea. It seems that Cramer's ideas must be interpreted in light of the parallel universes theory. All futures return the message, not just the best-chance future. There are more futures "listening" to the broadcast than just the one with the most sensitive and powerful receiver. In other words, each parallel world contains a single future event that connects with the present event through the modulation effect. Indeed, this is how the parallel worlds become separate from each other—once a modulation takes place, the parallel worlds split off and no longer interfere with each other.

What about time? If both the *possibility*-wave and the complex-conjugate wave are real, time must not be like a one-way river after all. Events that have passed must still be around. Events that will be must exist like new scenes around blind corners on the roads of life. And if both the future and the past exist, then, quantum physics implies, devices must be feasible that can enable us to tune in on the future and resonate with the past.

These devices seem to be our own brains, with our minds the controlling factors. When we remember a past event, we are not digging through anything like a file or computer memory bank. Rather, following quantum rules, we are constructing a past based on the multiplication of two clashing time-order streams of *possibility*-waves. Taken literally, this means that the past stream (the one flowing from past to present) must originate in the past the same way that the present stream (the one flowing from present to past) originates in the present. Past and present, then, somehow exist side-by-side.

It follows that the future, too, exists side-by-side with the present and that at this moment we are sending *possibility*-waves in that direction. Moreover, someone called "me" in the future is also sending back through time conjugate *possibility*-waves, which will clash with the waves being generated now.

If these streams "match," in the sense that the modulation produces a combined wave of some strength, and if there is a "resonance," meaning that the future and the present events are meaningful for me, then a real future is created from my present point of view and a real memory of sequences is created in the future. If the streams do not match—meaning that the combined wave is weak and there is no resonance—then the connection of that future and the present will be less meaningful. *Meaningful* here refers to the probability-curve. The implied law of time travel is: The greater the probability, the more meaningful the transaction and the greater the chance of it occurring.

The closer in time the sources of these waves are, the more likely it is that the two counter-time *possibility*-wave streams will produce a strong probability with a good chance of becoming real. Quite possibly, visionaries are those who successfully marry streams coming from time-distant sources, and people unable to cope with life are those who lack this ability for even the shortest time distances.

For most of us, though we might not be aware of it, time travel toward both the very near future and the immediate past already occurs in our minds. We saw the evidence for this as presented by Ben Libet and his associates, in chapter 4. Libet showed that we become aware of a bodily sensation, such as a sound that just happened, by referring back in time from a later moment of a brain signal arrival to the earlier moment of the bodily sensation. In other words, we seem to be aware of events before our brain registers them.

Think for a moment of the past, present, and future existing side-by-side. If we were able totally to "marry" corresponding times in each and every moment of our time-bound existences, there would indeed be no sense of time for us. We would all realize the timeless state that many spiritual traditions take to be our true and basic state of being. Instead, we find ourselves entering into one or the other parallel universe and thus failing to discriminate between the many past- and future-sending stations and all of the parallel universes attempting to communicate with us.

Thus we live time-bound lives disconnected to some extent from other possible pasts and futures.

What can we do to pick up a better, or perhaps different, signal from the future? If parallel futures are out there broadcasting back in time, surely there are some people who "hear" or "see" them. Perhaps among them are people who have lucid dreams. Perhaps certain mental disorders produce visions of the future. Even flying saucer sightings and "on-board" visitations with alien beings may be more than hoaxes or delusions. Perhaps the people experiencing them have traveled to a parallel world and back. Those we call visionaries may well be those who are able to tune out everyday life and tune in to these other worlds. "Past" and "future" are simply reference points based on our sense of now. Both are simultaneous in the parallel-worlds view of time. The specific past and future that we remember and appraise as real are simply those time-wave clashes that have the greatest strengths and the most resonances. Similarly, we can define "now" as the event, or sequence of adjacent events, that is the most meaningfully connected time-wave clash—the strongest clashes of waves that are "in tune" with each other.

Reality as we perceive it, according to quantum physics, depends on the subtle relationship between a *possibility*-wave and a probability-curve. *Possibility*-waves determine when and with what likelihood events occur. They don't do so directly, however, for they are submerged under the reality we perceive. Yet they are capable of both reinforcing and canceling each other, thereby affecting what we perceive by "shaving the odds." These odds show up as probability-curves, which determine the probabilities of the events in question. Probability-curves arise when two related *possibility*-waves multiply each other. We can envision one of the waves as moving forward in time between two events happening at different times, and the other as moving backward through time between the same two events. Through this process, time itself emerges, as do our immediate experiences. The causal relationships we see between events themselves arise from this deeper order where the *possibility*-waves reside.

TIME, MIND, *and* PROBABILITY

The happiest people spend much time in a state of flow, the state in which people are so involved in an activity that nothing else seems to matter; the experience itself is so enjoyable that people will do it even at great cost, for the sheer sake of doing it.

—Mihaly Csikzentmihalyi

As we have seen, *possibility*-waves and probability-curves, although related, have very different presences. Probability-curves appear sensible and directly related to our immediate experiences, while *possibility*-waves seem mysterious and one step removed. They affect us from some deeper reality—a sub-spacetime realm reminiscent of what we mean by the deep mind or the unconscious. We seem to have no control over these *possibility*-waves, which remain submerged below our levels of perception.

Probability-curves, on the other hand, we can literally count on, and indeed we do so every day of our lives. We need to know how to do things, and the doing of things requires making attempts and learning from them what works and what doesn't. Said differently, we assess the probabilities of success. As we learn to control the probability-curves of our lives, we begin to see the world in terms of cause and effect—certain actions will most likely produce certain outcomes.

Let me give you an example of the real-world presence of a probability-curve. Barring predawn emergencies, telephone calls, and doorbell rings, I know I can wake up naturally each morning at more or less the same time, about 7:15 A.M. Just to be sure, I usually set the alarm clock accordingly. But suppose I don't use the alarm for three hundred days. When I wake up each morning during that period, I look at the clock on the nightstand and check the time. Each possible time of awakening within certain temporal boundaries has a certain probability of occurring. Most of the time I wake up without an alarm at 7:15, plus or minus a few minutes. I only very rarely awaken as early as 7:00 or as late as 7:30. Even without carefully recording the data, I know experientially that the probability for my awakening peaks at 7:15 and falls off to no probability at all at times earlier than 7:00 and later than 7:30. In other words, it looks much like the well-known bell-shaped curve, shown in figure 9.1.

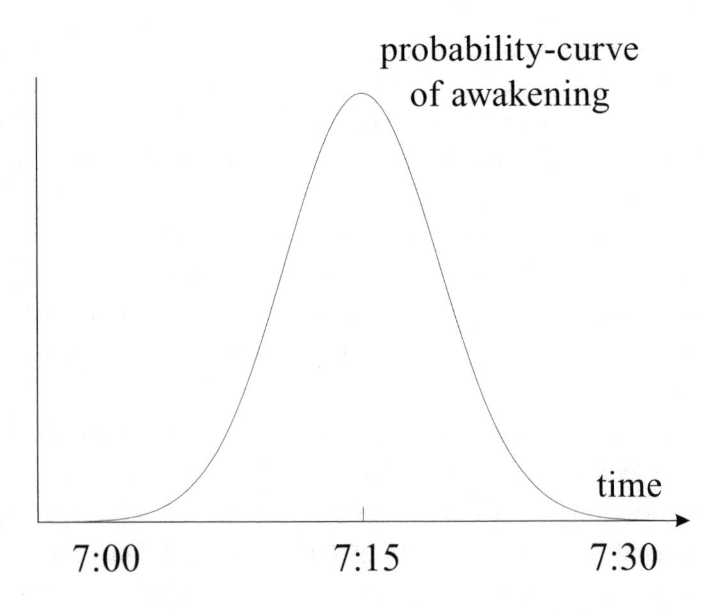

Figure 9.1. *The probability-curve of life.*

Let's suppose I want to make a daily record of my awakening times during this three-hundred-day period. All I have to do is write down the time I awaken each morning, recording it in, say, two-minute intervals; that is, I note the number of times I awoke between 7:00 and 7:02, the number of times between 7:02 and 7:04, and so on. No doubt I'll find that the number of times I woke up between 7:14 and 7:16 was significantly greater than the number of times I awoke between 7:00 and 7:02, or between 7:28 and 7:30. In fact, based on the curve, I can even predict that I'll awaken as follows:

60 times between 7:14 and 7:16
52 times between 7:12 and 7:14 (also between 7:16 and 7:18)
37 times between 7:10 and 7:12 (also between 7:18 and 7:20)
20 times between 7:08 and 7:10 (also between 7:20 and 7:22)
8 times between 7:06 and 7:08 (also between 7:22 and 7:24)
2 times between 7:04 and 7:06 (also between 7:24 and 7:26)
1 time between 7:02 and 7:04 (also between 7:26 and 7:28)
0 times between 7:00 and 7:02 (also between 7:28 and 7:30)

Note that the probability-curve doesn't allow me to predict which days I'll awaken between 7:14 and 7:16, but it will let me determine approximately how many days I'll be able to wake up at precisely within those times. Generally, probability-curves can't enable you to determine precisely what will occur on any given occasion. However, they do govern your behavior in the long run.

Consider another example: Take an open jar and put 30 quarters in it; shake them up and spill them out on a table; then count the number of heads you see. Repeat this procedure 300 times. It shouldn't surprise you that the number of times you count a prescribed number of heads showing can also be represented by the same bell-shaped curve. For example, count the number of times you see 14 heads. Add to this the number of times you see 15 heads and the number of times you see 16 heads. You'll find the total comes to 124 instances. In this manner you should find:

In 124 instances, the number of heads-up will be 14, 15, or 16.
In 73 instances, the number of heads-up will be 11, 12, or 13 (also 17, 18, or 19).
In 14 instances, the number of heads-up will be 8, 9, or 10 (also 20, 21, or 22).
In 1 instance, the number of heads-up will be 5, 6, or 7 (also 23, 24, or 25).
In 0 instances, the number of heads-up will be 0, 1, 2, 3, or 4 (or 26, 27, 28, 29, or 30).

When you do these kinds of experiments, you might be amazed that the results conform so well to the probability-curve in figure 9.1 and to the predictions I have indicated. In fact, very often gamblers who don't understand what a probability-curve really represents use the curve as a kind of predicting device. They assume that somehow nature is keeping track. For example, if after 250 tosses of the quarters, they had seen that the number of heads had been between 14 and 16 fewer than 10 times (whereas in 250 tosses the probability-curve predicts around 104 times), they would believe that in the next 50 tosses they should see between 14 and 16 heads showing up nearly half the time. However, nature is not keeping track and doesn't care how many heads and tails have shown up previously. A probability-curve does not, in fact, represent or predict any cause and effect between events. It only gives the odds of what will happen next, regardless of what happened before.

Not understanding this, the gamblers may be in for a surprise. If the outcome is already that skewed, most likely the coins are unfairly weighted—they have been altered so that the probability of showing tails occurs around 70 percent of the time, producing a skew in the expected results. Indeed, based on the results that the number of heads appearing 14, 15, or 16 times occurred less than 10 times in 250 tosses, one could even predict that the gambler running the game was cheating by using unfairly weighted coins. Hence, we come to rely on probability-curves and believe in them so much that we even invest our money on their expected behaviors and rue the day that we find ourselves cheated by unscrupulous investment companies who manipulate the odds in their favor.

How real are probability-curves? Certainly a probability-curve is not a physical object, so it doesn't exist "out there" as your alarm clock does. But it does represent real alternatives. We can always construct a physical representation of a probability-curve in the form of numerical data or a graph by performing the same procedure over and over again. Or we can gather up a number of identical or nearly identical people and have them all perform the same procedure. In all cases, the probability-curve will actually show up in our data.[1]

Probability-curves also have something to do with our awareness over time and even our happiness, given that happiness often is a reflection of how well we are able to exert ourselves in cause-effect relationships with the world around us. Even though we may do many things unconsciously, such as drive cars or wake up at a regular time in the morning, none of these skills are acquired without the necessary initial attention and gradual mastery of certain probability-curves while we are learning them. In fact, I can't think of anything we humans do that does not involve a probability-curve as we learn to do it.[2] Every skill we perform—even the ability to sit in a chair or read a book—implies a probability-curve in our consciousness whose effect is being expressed through time, even if we have forgotten all about working with that curve or never even knew such a curve existed.

When we no longer have to pay attention to the probability-curve involved in any skill, we label that skill a habit. Even though habits appear to be unconscious, they actually are not. We can become aware of them at any time—in many cases, simply by willing our minds to do so. Although habits are difficult to change, we all know they can be changed if we bring them into our immediate awareness. In fact, that is the practice of mind yoga and of yoga in general.

In essence, mind yoga is at first the practice of bringing habits to consciousness. This practice lays the groundwork for the kind of time travel I discussed in chapter 7, based on what we learned from the sphere of many radii. The time traveler sitting in the sphere travels in time as the sphere shifts his *possibility*-wave. Mind yoga enables you to make similar shifts in your *possibility*-wave. Hence we need to deal with the *possibility*-waves that underlie our habitual behavior and see how they get changed into probability-curves. The *possibility*-waves, however, are not available to us in spacetime. We must deal with them in the subtler level of reality that I call sub-spacetime and realize that our conscious experiences in fact arise from this realm. The question is, how do we access sub-spacetime?

Not in Space and Time

It may seem impossible that your experience of reading these words at this moment might *not* arise from the framework of space and time; nevertheless, it is true. We have come to believe that all human experiences are rooted in the physical world—that because the world we experience is physical, or material, our experiences are real. In other words, we have been taught that conscious experience arises because material objects move about and interact, producing changes in certain human body tissues, such as muscles and the nervous system.

You may be amazed when I tell you that there's no proof to back up this conclusion, in spite of the onslaught of evidence of muscle activity and brain activity that supports it. You may be scratching your head and wondering what I could possibly be referring to here. I am pointing to the one experience about which you are the most certain—your awareness of being in a body at this moment. That awareness is at times quite strong; for example, when you are learning a new task, such as a new dance step, or rehearsing a play, which means you must consider where and how you move about on the stage. Indeed, acting technique often asks the actor to immerse himself in his role, seeming to lose touch with who he is in order to take on the persona of the character he is playing.

At this moment, you are, as it were, playing a role. The character you portray may believe that he or she is fixed in his or her ways, not able to change one facet of the personality or modify a single desire or disgust. Yet most of this behavior, including the likes and dislikes, has been learned—may I say, rehearsed diligently—possibly over many long years.

A lot of rehearsing went on during the crucial early years from birth to the age of walking. Many more rehearsals took place as you grew from a toddler to a young adult. The teenage experience and its often painful conflicts prompted even more rehearsals, and in your attempt to belong to a group or association of friends, you acted out—possibly even against your better wishes—traits and beliefs you later grew to regret. With adult life and its responsibilities, these well-practiced character traits became second nature, so much so that you believed you knew who you were—you knew what was real about you, and you knew what was not.

Crucial throughout your development was your instinctive concern with survival—your ability to move your body and care for and protect it. Hence you have come to believe you have no choice in the matter: You are what your body tells you it is through the feedback loops in the nervous system.

The spiritual traditions of the world tell us that we are more than merely what our bodies tell us we are. And many of us—mystics and nonmystics alike—may catch glimpses of what they're talking about from time to time. But these insights also tend to be explained in physical terms. The conviction that material objects are more real is so convincing that it becomes our default assumption. It takes a lot of teaching and practice—a steady practice of mind yoga—to convince us otherwise.

This habitual notion—that the physical world is more real—continues to make its presence felt in science, even though quantum physics tells us repeatedly that the basis of this notion is flawed. That basis is the conviction that the world we experience is the world of classical, Newtonian physics, and it is fixed in our minds through powerful logic and buttressed with mathematical argument. If this view of things were true, it would mean that consciousness must arise out of the interactions of material objects. But in none of our neurological or biological studies of these interactions do we see anything like consciousness emerge.

It is important to note that when we observe consciousness in others, we are objectifying it, seeing it in terms of what our sense organs tell us. But when we observe consciousness within ourselves, this process of objectification totally, and necessarily, fails. The faulty assumption is first of all the belief, based on our observations, that others we see around us are physical beings who "have" consciousness; and secondly, that the connection between material processes and the consciousness we think we're seeing applies to our own inner experiences in the same way. Hence we believe that, whatever mind is, it has grown out of matter in some causal manner.

Let me go through the evidence for believing that mind arises out of matter. It is strongly based on the theory of evolution.

We see that intelligent entities like animals and people exhibit varying degrees of complexity. We appear to be more complex than single-celled animals, for example, so we take it for granted that our sophisticated minds arose from this complexity. In brief, amoebae don't think because they are too simple; we think because we are sufficiently complex.

We understand that complexity arose through the forces of evolution. We see evidence that through random events affecting genetic material of simple animals exposed to environmental changes over many life spans, these life forms adapt. They learn to make changes that fit the changes occurring in their environment. In accordance with the law of the "survival of the fittest," those that don't adapt die out and no longer reproduce themselves.

We understand survival in terms of the law of cause and effect. How a creature adapts itself depends on causes affecting its life span handed down to it from its ancestors. Today we seek to find the genetic markers in human beings that are supposed to be able to determine if an individual is susceptible to colon cancer, diabetes, or any number of modern diseases plaguing Western and other well-nourished societies. (Often societies that are poorer materially find themselves dying from hunger, bad water, and impoverished living quarters.)

These observations—and I am sure more could be added—would certainly be true, provided the ground on which they are based was solid itself. But even though these arguments are scientifically based, they still rely on the commonsense view that space, time, and matter are fundamental. Hence, anything that doesn't fit into a matrix built up from these three elements, must not and cannot be real or have any effect on reality. But quantum physics tells us relentlessly that there is something prior to space, time, and matter. I call it a sub-spacetime. Others have called it the imaginal realm, and in present quantum theory it is posited to be an infinitely dimensional space. Quantum processes are vital in this realm, and what we call consciousness appears to play a fundamental role at the level of even the most primary matter, consisting of atoms and subatomic particles. With the presence of consciousness at that level and with principles of quantum physics factored in, modern science necessarily changes the belief expressed above that consciousness or mind arose from matter, as the theory of evolution would have it. This requires a careful look at what quantum physics tells us about the relationship between *possibility*-waves and probability-curves.

The Role of Mind in a Sub-spacetime Realm

Quantum physics deals with a sub-spacetime world that is beyond both matter and mind. This world represents possibilities that appear as waves upon which the mind plays a role in the construction of reality. You can think of the sub-spacetime realm as the great unconscious Mind of God or the fundamental ground upon which reality appearing as mind and matter emerge. The way in which this all takes place can be grasped through the roles played by *possibility*-waves and their transformation into probability-curves.

Possibility-waves appear to exist purely within the sub-spacetime realm. Probability-curves form our "out there" consciousness; they mark time and bind the mind to the present. In contrast, *possibility*-waves form our "in here" consciousness, free the mind from the present, and allow the mind to be free of time.

Unlike the proverbial fish that can't imagine the ocean in which it swims, the mind appears to be capable of conceptualizing the sub-spacetime realm in which it exists. Through mathematics and science as well as yoga and other disciplines, the mind can grasp and make meaning out of the intangible realm of existence that arises prior to space, time, and matter. Our ability to do so, to imagine beyond the immediate sensory experiences taking place within spacetime, is truly remarkable and in fact very mysterious. It seems to be the way mind enters into the sub-spacetime realm, indeed "plays" within it and emerges from it with the whole material universe in tow.

The mind has the facility to form the "out there" material world of space and time. I am referring here to the operation of "squaring"—the multiplying of one *possibility*-wave by its complex conjugate *possibility*-wave. By this squaring mechanism, objective realities (probability-curves) are created "out there" from the *possibility*-waves undulating within the subjective-unconscious-imaginal sub-spacetime "in-here" realm. In the process of squaring and then dealing with probability-curves, the mind moves from the purely imaginal realm into awareness of the physical realm.

How does the mind construct reality? I speculate that it is by accessing the squaring operation, which I believe is what yoga and other mental disciplines show us how to do. There are two processes involved in the squaring operation. "Squaring," which is an act bringing the mind to a focus, and "unsquaring," which allows the mind to defocus or "let go" of whatever it has been focused on.

How do these mind actions compare to yoga teaching? Teachers of yoga tell me that the purpose of poses (*asanas*) in yoga is to bring awareness of body to mind. The way in which this awareness occurs takes place in steps that often involve holding a pose for a period of time and then "breathing into it" and adjusting the pose accordingly. Mind yoga is really no different; the language has just changed to take into account what we know about the sub-spacetime realm and its relation to the physical spacetime world. When the mind "squares" a *possibility*-wave, awareness arises or manifests as something physical; an objective comes into focus. In yogic terms, this could be the awareness of the body in a pose or the awareness that something new is emerging in thought or feeling. So "squaring" is the same as entering into a pose in yoga. When the "breathing into it" takes place, the pose is adjusted. This is the same as "letting go" or the process of "unsquaring," which changes the next "squaring" operation by making it either more or less probable of success. Letting go introduces indeterminacy into the next outcome. It means that the next act of focusing (squaring) will not be as predictable as the focusing act prior to it apparently was. In this way we are able to make adjustments, just as the yoga teacher adjusts your pose in a class. If you held onto the pose, you wouldn't be able to change.

We could say that our potential ability to "square things up," meaning address the way we handle our likes and dislikes and manage our thoughts and feelings "in here," lies in being able to align our thinking either with the past or with the future. Practices like meditation and yoga offer us the prospect of adjusting the dynamics involved in the squaring and unsquaring operations, opening the door to new possibilities "out there," and, most important to our time travel discussion, being able to move a point of view either backward or forward in time. At first the idea of being able to adjust the squaring operation may sound as strange as the concept of the squaring operation itself. It turns out, however, that the squaring and, with it, the possibility of alteration through unsquaring fall within your mental abilities—your field of consciousness. Indeed, if they didn't, you wouldn't be able to make up your mind or change it. For squaring and unsquaring are what we do when we exercise the mind to form a point of view or open the mind to learn something new.

You can think of a point of view in any given moment as a *focal point*—a site where a *possibility*-wave from the past or the future meets and mixes with a *possibility*-wave issuing from the present moment, thus squaring and producing a probability-curve. We saw with the bell-shaped curve that many individual events go into constructing a single skill of ours such that it becomes a habit. The same applies to our construction of reality. Any single focal point, or moment of reality, may be more or less precise—just as any given morning I might wake up closer or farther from the predominant time of 7:15 A.M. The more focused moments are when we perceive reality as objective, "out there." The more unfocused or blurry moments are moments of subjective perception, directed "in here"—or else directed "out there" but with less insistence that the world presents itself to us as we expect it.

Focused and unfocused sites occur in a sequential pattern. Our conscious experience consists of a sequence of these focal points, sites of more specific focus separated by sites of unfocused possibilities.

Continuing with my speculative point of view, I suggest that this sequence of focused and unfocused sites is what offers us the perception of time. The sequence provides us with a temporal sense—the ability to objectify our experiences through the measure of time, as well as our own self-awareness and memory. In this sense, time and consciousness are actually different labels for one and the same thing—the process by which we become attached to material existence. Having recognized what causes our attachment to the everyday world of matter and causality, we can detach ourselves from its confines—that is, we can defeat time.

Figure 9.2 is a schematic representation of a possible sequence of focused and unfocused points.[3] These represent moments of perception of an object as it passes through various

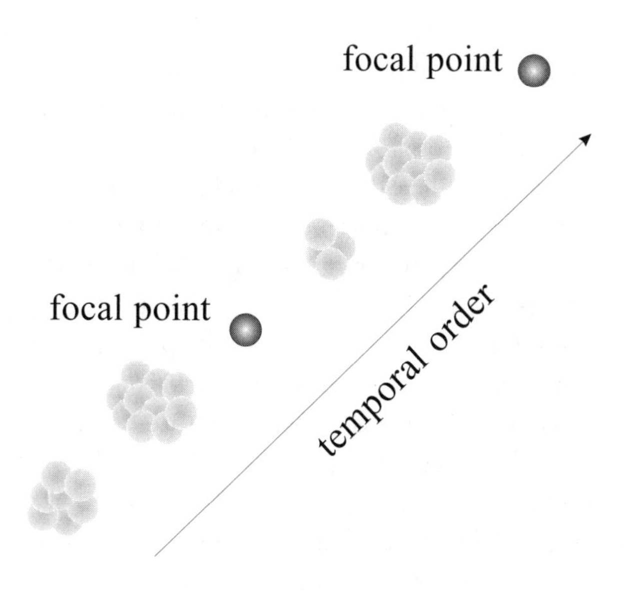

Figure 9.2. *Focal points of consciousness mark time.*

focal changes. You can think of these as a series of "snapshots" depicting the *possibility*-wave of the same object taken at different times. Sometimes it is more blurred, sometimes more focused. The blurring in a single snapshot represents the uncertainty in the possible location for the object in that moment. The wider the blur area in the diagram, the smaller is the probability for locating the object at any one place within the blur area—that is, the blurrier, or less certain, my conscious perception of the object. On the other hand, the narrower the blur area, the greater is the probability for locating the object at any place within the blur area and the clearer and more objective the object appears for me.

However, the blur is not due to any inadequacy on my part or the part of any observer to locate the object. The object appears blurry because it has no actual fixed position. That is, in fact, the whole point of quantum physics—to deal with the blurs and find some way to take them into account. One way is simply to deny that the object exists at all until the blur becomes focused. This was the Bohr interpretation discussed earlier. Another way is to think of the blur as if it were an overlap of possible "real" objects—each object existing in a separate parallel universe—just as if one had each possible object appearing as a picture on a transparency and overlaid them above a light source. This was the parallel-universes interpretation discussed earlier. The greater the blur area, the greater is the number of parallel universes involved. These potential parallel universes exist in sub-spacetime. Physicist Amit Goswami refers to this sub-spacetime array of cellophane universes as a *potentia* or *idealistic* reality that resides beyond spacetime.[4] Physicist Roger Penrose wrote extensively about this idealistic reality as a third world of existence, akin to Plato's world of ideals.[5] Penrose takes the existence of this world quite seriously and believes that consciousness works in a nonalgorithmic manner in its actions of squaring. As we have seen, I take it that the squaring arises from the mechanism suggested by Cramer's transactional interpretation.[6]

Let's see if we can discover how objects manifest in spacetime from beyond spacetime, using this picture as our guide. We can imagine a sequence of blurry and focused photos as shown in figure 9.2. We also note that the sequence can be put into an order. Which "photo" should be first and which second? In other words, how should we sort the sequence?

To deal with the sorting problem, think about what it would mean if no focal points were to occur in the whole universe. If we could step back from it and watch, without squaring, all *possibility*-waves, we would see the universe evolve by getting more and more blurry. In fact, all that would exist would be a growing blur of uncertainty with nothing really ever happening and no one really ever witnessing anything occurring. If such a thing were to occur in the universe around us, it would become more and more chaotic and entropy would reign supreme.

But nothing like that occurs because mind enters into the growing disorder and plays a role by bringing in a focal point of view—an ability to choose—which it does through the act of squaring. However, in so doing, although an orderly world of cause and effect arises, a quality of life begins to fade. For with greater certainty, there also comes less joy, less spontaneity, and certainly little room for mystery and surprise. The dance of life appears to be a dance involving focusing and unfocusing, squaring and unsquaring—letting go of—*possibility*-waves, so that mastery of life and the occurrence of novelty, joy, bliss, and sorrow can continue.

Taking into account the continual play of mind within the sub-spacetime arena, in fact, there seems to be a natural order of relative focus and blurriness—certainty and uncertainty—to the way we construct the sequence of focus and unfocused sites we call reality.

Any sequence of three sub-spacetime sites containing a focal point is called a "triplet." In any such sequence, *the normal and natural order is a larger blur prior to the focal point and a smaller blur following it* (see figure 9.3). In other words, a focused site of consciousness is preceded by an unfocused site of greater possibility or uncertainty and is followed by an unfocused site that is nevertheless more certain than the previous unfocused site. Said yet another way, the relative certainty of a focused site reduces the possibilities. We know little or nothing about the object—or a group of objects or a whole scene—before the focused perception, and we know more about it afterwards.

I call this the *quantum law of normal time order.* The word *normal* is important. This is the order that nature, including the human mind, follows on its own. And in so doing, nature creates the objective time order we have grown accustomed to.

Normally, in relation to the focal point, the precedent blur appears larger than the posterior blur. But it is also possible for both blurs to be the same size, or for the precedent blur to be smaller than the posterior blur. Each sequence of three points—each *triplet*—depicts a different order, with the focal point always in the middle of the blurs. If the precedent blur is larger than the posterior blur—the "normal" situation—the sequence represents some gain of control over and knowledge about the object in question. If the precedent blur is the same size as the posterior blur—the "balanced" situation—the sequence represents some habitual behavior and a degree of control over the object. If the precedent blur is smaller than the posterior blur—the "reversed" situation—the sequence represents some loss of control over and knowledge of the object.

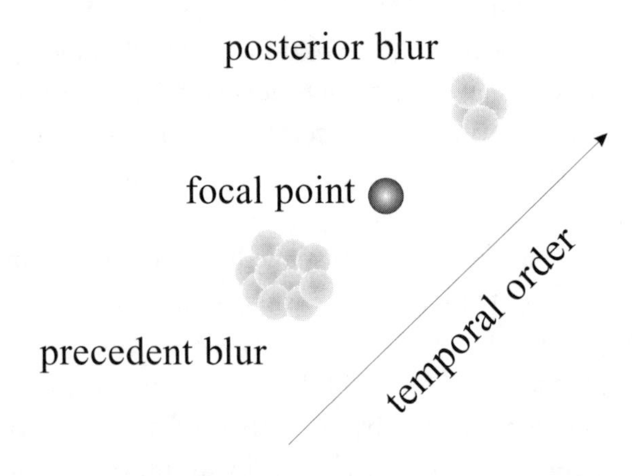

Figure 9.3. *The quantum law of normal time order: Small blurs follow big blurs.*

You can think of the focal points as places in spacetime where objective awareness occurs and the more blurred sites as places where objective awareness diminishes but subjective awareness persists. Squaring, which changes a *possibility*-wave into a probability-curve, corresponds to the process that produces a focused point, or "sudden awareness," while unsquaring produces a blurred site, that is, "unawareness" or unconsciousness. That is to say, blurring represents a fuzziness in the sense of reality or the content of the perception. Squaring, it should be mentioned, doesn't always produce perfect focus or control. Some fuzziness could remain, depending on the form of the *possibility*-wave and the degree of concentration or desire.

Before squaring, multiple possibilities exist in sub-spacetime. After squaring, the number of possibilities is less. In other words, even though you might not know where the object is, it has now become tangible. Although the object is in focus in sub-spacetime, its actual position in the objective realm of spacetime still remains undetermined. You can think of it as having a real existence somewhere at sometime. In other words, it counts as an objective choice that is present, even though you may not actually observe that choice. It exists in the same sense that different choices exist as possible outcomes. This means it has entered what I'll call objective consciousness, even though you may not yet experience that knowledge.

As an example, suppose someone flips a coin. At this point, your degree of uncertainty about the coin is maximal; while it is going through its gyrations, its *possibility*-wave for heads and tails continues to change. But then the coin lands, and the person who flipped it observes which side is up, but doesn't tell you. Now you have more certainty about the coin: You know exactly where it is, and you know that it has been observed and lies with either heads or tails up. Even though you still don't know everything about the coin—i.e., which side of it is visible—you know more about it than you did before. For you, it has shifted from the realm of a *possibility*-wave to the world of probability-curves.

In this example, the gyrating coin corresponds to the precedent blur. The observation of the coin corresponds to a focus; and the coin after it has landed corresponds to the posterior blur, which is more focused than when the coin was in the air.

MIND YOGA AND TIME TRAVEL

Mind yoga consists of controlling sequences of focal points and blurs, in other words, of controlling triplets. To repeat, a triplet is a sequence containing two blurs and a focal point. All of our conscious experiences in life arise as chains of triplets. These triplets evolve through our efforts. Each time a triplet evolves, we gain (or lose) some control over our bodies and our environments and learn how and to what extent we can manipulate them. We learn not only how to create points of view but how to connect them, to construct a chain of causality by which we begin to see the past as a cause for the present and the present as an anticipation of the future.

In any triplet, we mark the past and the future relative to the focus. (In figure 9.3 above, we see such a sequence making up what I call the natural law of temporal order, wherein the larger precedent blur occurs before the focal point and the smaller posterior blur occurs after it.) Now compare a particular triplet with any other. To control the object, you need an evolution from a less to a more focused triplet. By comparing the sequences, you decide which is which—which triplet belongs in the past and which in the fu-ture. In other words, a rule of comparison arises such that what you experience as past and as future conforms to a consistent rule of or-der we call causality. Causality gives us our sense of control over life.

In figure 9.4, we see a series of triple sequences arranged according to a diminishing of the blurs until a triplet arises where the posterior and precedent blurs are approximately the same size. This sequencing represents a growing control over the object and marks habitual behavior; that is, the degree of uncertainty both before and after the actual perception of reality is minimized and hardly changes. When control is established, temporal order or causality emerges. We gain an understanding of the process and the ability to control and predict its behavior. In other words, the sequence becomes a habit wherein the probable consequences are narrowed, as indicated by the narrowing of the blurs and the emergence of a greater number of focal points. In time, the unfocused past becomes entirely unavailable and only the focused control remains as past reality. What we believe to be the past, whether pleasant or not, appears to us as events over which we had some control. Even though we may have had a lot less control over the actual events, the fact that we remember them itself gives us a certain degree of control over them.

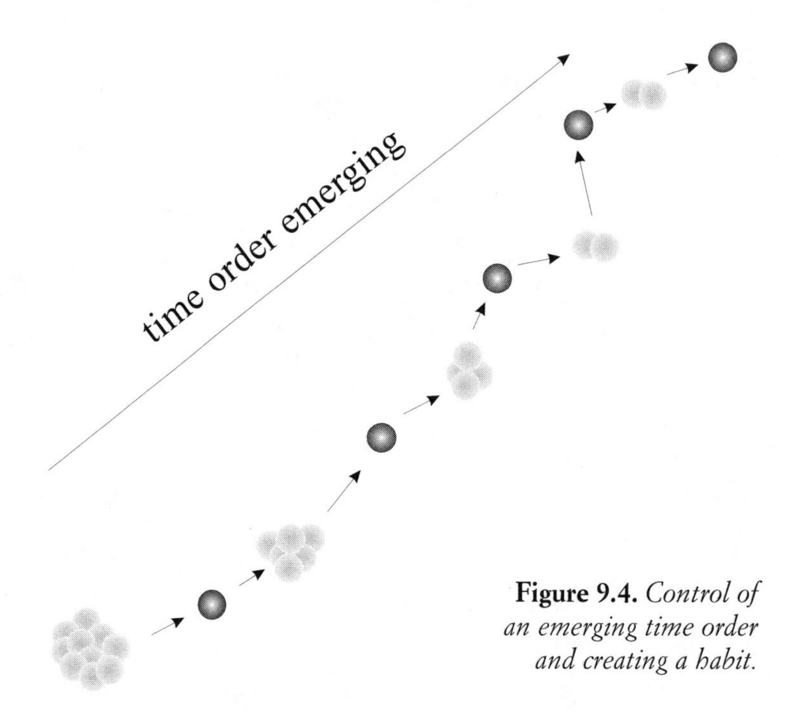

Figure 9.4. *Control of an emerging time order and creating a habit.*

The natural course of *possibility*-waves, without the intervention of consciousness to create a focused point, is to go from a more- to a less-focused pattern. That is why our perception of reality tends to blur and spread out after a focused moment. We can use this natural process to advantage. In figure 9.5 we see a "letting go" process wherein the object appears to be going in reverse of the picture shown in figure 9.4. This corresponds to letting go of any anticipation or control and also of any clear memory of the past, which is the key to many meditation practices and most particularly to yoga. When we let go, we "unfocus" or free up any picture we had of where, when, and how an object exists. We let go of expectations. Once an object is "freed," moreover, the possibilities associated with it increase at any particular location in spacetime. The unfocused object "spreads out," meaning that we are no longer looking for possible positions the object may have in the future, and no attempt is made to square again the *possibility*-wave for any of these possible positions. The spreading of a freed object continues and in general would continue to fill the

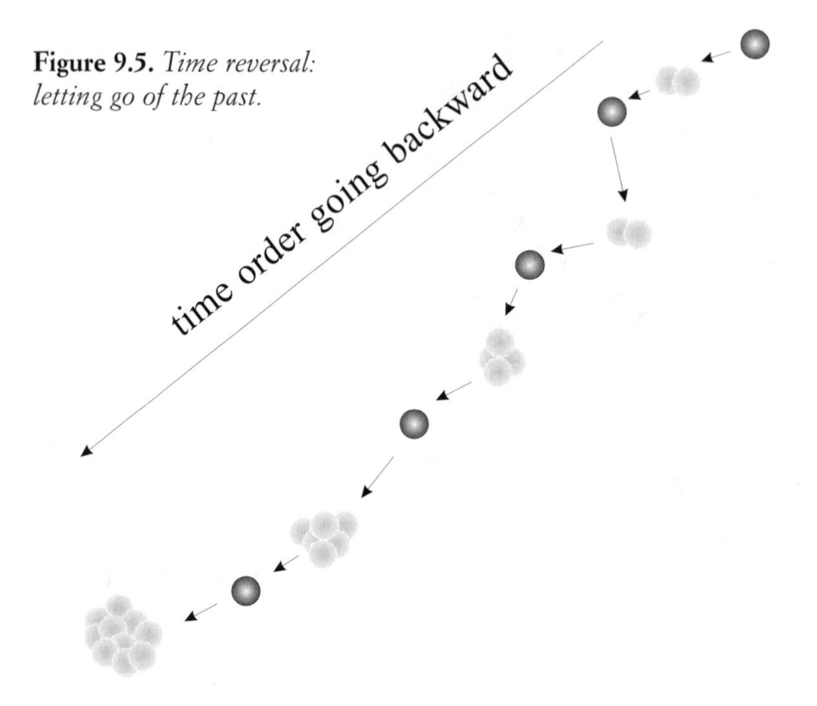

Figure 9.5. *Time reversal: letting go of the past.*

entire universe if no further squaring ever took place. In psychological terms, this corresponds to forgiving yourself or another, giving up expectations, and even facing the moment of death.

These two abilities, focusing and letting go, constitute the basic binary activity of conscious life. Through focus we learn to control or master the skills we need to cope, and through unfocusing we learn to relax and let the world in without judgment. The progression created by successive focal points of concentration and fuzziness—"crunch" and "relax"—establishes a time order, which is the basis for the history of the individual.

From your birth until your death, you will believe, and the world around you will most likely insist, that you have lived your life along a time line pointing to ever-increasing years and old age. But as we have learned, that line has several loops in it, allowing you through the processes of squaring and unsquaring to move forward or backward in time while the remaining world moves on in agreed temporal order. Becoming a time traveler through mind yoga means learning how and when to focus and unfocus your mind. Traveling forward in time means following the natural order shown in figure 9.4, while traveling backward means reversing that order, as shown in figure 9.5.

It is interesting to compare the evolving time order presented here and its reversal with the thermodynamic arrow of time introduced in chapter 4. There we saw that, contrary to what I am proposing here, the requirement of energy losses due to friction inherent in any machine's operation tends to imply that the direction of time is associated with these losses in available energy to do work. In today's technology, we look at this as the law of increasing entropy and see it as a measure of these energy losses. We tend to believe that entropy increases as time goes on. Hence the more chaotic it becomes the later it becomes.

I explained that quantum physics indicates that without observers, *possibility*-waves tend to spread out emulating this law of entropy. But no time would appear in such a universe. Possibilities would just go on increasing without end. But, and this is crucial, the world involving ourselves appears to reverse that law. When such a reversal arises, we usually give arguments based on energy to account for the reversal, in the same sense that a refrigerator reverses the law of entropy by pumping heat from a cold body to a warmer body simply because we add energy to it. Consciousness acts in the universe like the energy added to a refrigerator—it reverses the law of entropy. And because it is so prevalent, I take this to be the natural temporal law of the universe including mind, whereas thermodynamics, physics, and science in general omit the actions of mind in their calculations.

In summary, when an unsquaring operation occurs, a letting go or some kind of release takes place, and for the system time goes backward. For example, you experienced this state all the time as a baby, when you were constantly opening your mind to learn something new in order to understand your environment and grow up. The letting go that takes place in mind yoga is akin to returning to this childlike state. Instead of maximizing entropy, which implies decrepitude, it is a simpler, more probable, and thereby more creative state, and while in it, time travel becomes possible.

CHAPTER TEN

The SPIRITUAL DIMENSION *of* TIME TRAVEL

Time flies like an arrow.
Fruit flies like a banana.
—Groucho Marx

Like Groucho's pun on the words *flies like* in the above joke, we tend to use the words *mind* and *time* to mean different things without realizing some new and surprising ways in which they can relate. From what we have seen in chapters 8 and 9, we are now able to comprehend how the concepts of time and mind are reconcilable—and can, in fact, have the same meaning.

This realization comes from recognizing that the mind works by a process of defocusing and focusing, which means it can learn how to let go of memories (allowing them to return to the great blur of possibility), and it can learn how to focus in on possibilities (allowing a specific event to come into being). This process is also the practice of mind yoga. While that much may seem sensible, in chapter 9 I added a new twist by suggesting that this process is actually what brings time about. Hence, the relation between mind and time takes on a whole new meaning.

Through the practice of focusing and defocusing, although we may not realize it in our everyday lives, time is actually being created, and this creation makes time travel a necessary part of the way that mind functions and the way time works. Please review the illustrations in chapter 9: The mind would travel forward in time following the natural sequence shown in figure 9.4 and travel backward following the reverse sequence shown in figure 9.5. Every day of our mental life, we follow both directions as we go about doing our tasks. When we let go of old habits, we move backward in time; and when we follow habitual behavior, we move forward. Of course, this process refers to our *subjective* time-sense, which may run counter to or in the same direction as *objective* time.

Not only does our "in here" time direction change in the focusing and defocusing processes, but the rate the process ensues can also change. That is, we can move forward *or* backward through time either faster or slower than the rate at which objective time progresses. I discuss this rate change in more detail later. It is my speculation that, through our realization of the different way time functions "in here," we actually came to the discovery and agreement among our fellows needed to create objective time. Evolutionarily, we needed objective time because time travel was confusing. Human beings created objective time to assist in their communal survival.

I fully realize why many, including scientists, would object to the notion that time travel is a matter of the mind. One reason is simple: There is no way to argue or discuss the matter in terms of a repeatable experiment. How would we do so? When it comes to the technology of time travel (discussed in chapter 7), this objection probably doesn't arise. It is certainly difficult, perhaps impossible, to create the sphere of many radii in a laboratory. Nevertheless, the possibility holds a scientist's attention because if the device could be built, the experiment could be carried out—just gather together the necessary ingredients, follow the recipe, and *voila!* You have a time machine. Whether or not it works becomes a question of technical skill.

Time travel as envisioned through the mind, though, is a very different thing, and bringing it into the discussion immediately plunges us into the areas of psychology and spirituality. The mind is not objectifiable; hence, experiments dealing with it are ultimately indirect—including experimentation on human and animal behavior and its possible modification through physical means or stimuli of some kind. Using the mind for time travel is not only indirect, it is also highly subjective. As such, it becomes part of spiritual discipline as much as of psychology.

I briefly explained in chapter 1 that time travel using the mind gives the traveler a whole new vista of possible exploration, but it also requires a sacrifice. Many of us feel the need to carry lots of baggage when we travel any distance; however, to time travel we must leave some seemingly very important luggage behind. That baggage, which is our egos and our sense of individuality, may be something we would rather not and, indeed, may not even be able to leave. It is this luggage that prevents the time traveler's access to the future and the past. We have spent our lives developing our sense of who and what we believe we are. The ego appears to us as robust and indispensable, yet ancient spiritual teaching tells us that it is only an illusion through which each of us conceives of ourselves as a singular entity, or "I."

Just as all movies do, this illusion captures us, and we soon feel that we are no longer an audience witnessing but actual characters in the film. Coupled to this chimera, the animated picture inundates our senses so that we believe we live in an objectifiable world of time and space and that anything outside of this "reality" is purely subjective.

Perhaps Plato was referring to this phenomenon in his allegory of the cave, in which prisoners are chained so that they can only watch their shadows cast by firelight upon a wall, all the while erroneously believing that they *are,* in fact, those images. Would we readily accept the reality behind the images if we could? It would no doubt be quite difficult even to admit that we are "chained" by our desire to "exist" in material forms. Dropping those chains without some spiritual discipline could be quite devastating.

Chapter Ten

The Minds of Physicists

Some may think that, as a physicist, I am out of my depth dealing with what seem to be psychological, behavioristic, or spiritual phenomena. Before I dive deeper, let me counter this supposition with a simple statement. Possibly our notions about the mind have been put in the wrong categories to begin with. Because the mind cannot be objectified, objective—psychological or behavioristic—ways of thinking about it will not and cannot pertain. But a new, subjective model of the mind could change the way we think scientifically and influence research in a science of mind. It may even help us understand evolution in a new way, or determine how and why a self grows to believe that it is nothing but the body, even though it may suspect that there is something more developing.

This idea is not as radical as one might think. For example, today in physics we are currently at work on a whole new way of thinking about how the universe began and what it is made of. The new way is called *string theory,* and one of its outstanding and surprising features is that it is based on ideas that are unprovable by any experimental test. In a recent interview, Nobel Prize-winning physicist Sheldon Glashow noted:

> [In my day] experimenters and the theorists were in very close contact. This intimacy continued and it continues today certainly at my university. But oddly there has been a new development, in which a new class of physicists is doing physics, undeniably physics, but physics of a sort that does not relate to anything experimental. This new class is interested in experiment from a cultural but not a scientific point of view, because they have focused on questions that experiment cannot address. So this is a change. It's something that began to develop in the '80s, grew in the '90s, and today attracts many of the best and brightest physicists. It's called superstring theory and it is, so far as I can see, totally divorced from experiment or observation. If not totally divorced, pretty well divorced. They will deny that, these string theorists. They will say, "We predicted the existence

> of gravity." Well, I knew a lot about gravity before there were any string theorists, so I don't take that as a prediction. The string theorists have a theory that appears to be consistent and is very beautiful, very complex, and I don't understand it. It gives a quantum theory of gravity that appears to be consistent but doesn't make any other predictions. That is to say, there ain't no experiment that could be done nor is there any observation that could be made that would say, "You guys are wrong." The theory is safe, permanently safe. I ask you, is that a theory of physics or a philosophy?[1]

It seems we are seeing a new trend that leads physicists trying to grasp the meaning of the universe into the untestable, unobjective world of the imagination, which lies entirely in the subjective realm. I don't think this is a temporary detour on the way to grasping the meaning of the universe; rather, I propose it is a deep clue into the subjective underpinnings of how the universe works. In essence, there is no universe present without the imagination, no imagination without a mind, and no mind without consciousness.[2] Thus, as we begin to probe into the structure of consciousness, we come to its spiritual foundation in ancient wisdom. It is a spiritual world, after all, and understanding time travel will help us to see this reality more clearly.

The Spiritual Secret of Time Travel: Surrendering the Ego

Spirituality is a difficult subject to discuss. Perhaps it is even more difficult for physicists because of our objectivistic training. Being a physicist and a writer, I feel this difficulty keenly, and I confess that at times I find some people who profess "spiritual wisdom" pretentious if not downright fraudulent. Nevertheless, in the following discussion I want to speculate farther into this realm than I have done before. Hence, I will use phrases that may seem pretentious, such as "spiritual truth," "the Mind of God," "the true message of spirituality," and so on. Please accept them without attaching any egoistic concerns to them. I believe I see how time, mind, and the spiritual nature of humankind are deeply connected, and in order to discuss this connection I will need to enter into the territory usually and legitimately held by the clergy and scholars of religion. I am thus treading on shaky ground for a physicist. Let me put it as simply as I can:

The key to our grasp of the true message of spirituality, as well as of the means to time travel, is our ability to become aware of that extra baggage we all carry called the ego—or our normal, everyday state of waking consciousness in which we think of ourselves as "I." But what is the ego, really? I like to think of it as a closed surface in the imaginal realm of the mind. Any closed surface will do. You could imagine a spherical bubble, for example. For this discussion, think of a six-sided cubical room. Each of its four square walls and the ceiling and floor has its inner surface coated with a mirror. Hence, all light within the cubical room reflects inward to the cube's center. The room is the ego. It thus acts by protecting whoever sits inside of it from the "out there," and, because of the mirrors, it also acts as a focusing device. In so doing, the ego provides a kind of theater in the mind that keeps us entertained by this continuing reflected lightshow. The mirrors aren't quite perfect, so that light from "out there" gets in and mixes with the internal reflections. Hence, we are able to compare the information from "out there" with the lightshow going on "in here."

This lightshow holds our minds in our bodies in much the same way as we become glued to our seats when we are captivated by a good movie. Although we are only witnesses to life's passage, we experience that passage as participants. We see ourselves, not as images on a screen, but as real selves each inside of his or her own private world—our bodies.

Why does this happen? It seems to me that God requires a great number of focal centers in order to awaken from the illusory trap of material existence. Each center appears as a single self with the ego providing a surface of surrounding "mirrors" producing the focusing, as I discussed above and in chapter 9. In other words, God has become trapped in spacetime, fully aware that He or She would become so trapped and willing to have it happen. This is the great sacrifice talked about in ancient teaching. In order to make a physical world that would remain robust, God had to become part of it.

What I have discovered through the theories of quantum physics, parallel universes, the special theory of relativity, and the general theory of relativity has allowed me to understand how God becomes so self-trapped. In order to change possibilities of life into actualities, the Mind of God must anchor itself in spacetime and appear as individual minds within bodies. Put simply, by so doing, God pins Her or His Mind in time and appears as conscious bodies in space.

The world of matter may be an illusion, no doubt, according to many spiritual beliefs. But it certainly seems real to all of us who have apparently not remembered this sacrifice of God. As such, the illusion appears as the ongoing "movie" that connects objective time and subjective mind. This entrapment of the Mind of God not only stabilizes the universe (without Mind in the universe nothing material would ever appear, according to quantum physics), but it also enables Mind to experience itself as "other" beings. It provides a common awareness of the physical world, and it gives us a sense of objective time and space. Time becomes what we agree it is through common experiences.

Let me now look at how time travel is possible through surrender of the ego: Ego and time-boundedness are connected. What we think we are, and what we truly are, are not the same things. What we think we are is greatly influenced by the egoistic reflections of the mind. What we *are* remains as the Mind of God. The ego forms as a boundary separating the "out there" from the "in here." It does so as an evolutionary attempt to protect life. As life evolves, older schemes of protection are no longer needed (although it is not entirely clear which schemes should be dropped) and, by changing our self-imposed boundaries, we can escape from the spacetime matrix that binds the mind.

Time is intimately related to mind and thought and, as we saw in chapters 8 and 9, to possibility and probability; hence, time is as real as thought. However, we all have access to that timeless, spaceless realm where time itself is created. Time is a projection of mind, and by changing our ego structures—the way we think about our roles as individuals in the world—we can defeat our ego-conditioning and become aware of our ability to time travel.

This escape from ego-boundedness and consequent possibility of time travel provides great and singular benefits. Identified with our bodies, which are limited in space and time, we sentient beings have lost the experience (if it can be called that) of residing in timeless, spaceless eternity (now being investigated by physicists studying superstring theory and quantum physics, although they may not think of it in this manner). Time travel frees us from such limitations. This freedom typically occurs when we practice any spiritual discipline that alters the relation between body and mind, especially body yoga and mind yoga. It allows us to move through time and to improve our quality of life by revisiting past errors and learning to forgive ourselves and others. It gives a fresh, life-affirming quality to everyday existence. Perhaps (and maybe most important for some), it teaches us to slow our body's aging processes.

This may sound far-fetched. However, substantiation of these ideas can be found in the ancient practices of yoga and in the study of how physics shows time travel is a necessary part of its structure.[3] We may reasonably conclude that time travel is inherent in the universe, not a freaky sideshow that plays a little role. For example, consider the parallel between the sacred notion that time is circular (as discussed in chapter 2) and the notion in the general theory of relativity that gravity bends time into a circle or closed timelike line (as discussed in chapter 5). They may be saying the same thing from different points of view.

As I explained in chapters 7 and 8, there are two kinds of time travel: ordinary and extraordinary. The first requires material technology that has not been created to date. The second requires practice through mind yoga, which is readily available to us all. Those who practice extraordinary time travel will recognize its spiritual dimension. Those who wish to wait for the new technology to practice *ordinary* time travel, however, will probably nevertheless need to deal with the spiritual dimensions of *extraordinary* time travel. Ordinary time travelers will begin to see that they are not as separate from each other as they may have believed in consequence of obvious differences in culture and upbringing.

Along these lines, we are aware that different cultures have different time concepts and hence view time differently. Both Native Americans and Australian aboriginal people view time as a duality—a linear time that you and I are familiar with, plus a circular or sacred time. With the discovery of curved time in the general theory of relativity, we may be seeing reconciliation between the two views. Curved time comes about through grav-ity—the force that holds the earth together and all objects together. We can think of gravity, as well as matter and energy, as a sacred part of nature. I see this as a strong indication that physics and spirituality are indeed realizing the same truth.

Ancient spiritual wisdom also suggests it is possible to make time speed up or slow down by using the mind. These interesting possibilities can be considered from the views both of physics and of ancient wisdom. In chapter 8, we discussed the idea of using the mind to time travel in connection with the focusing and defocusing of mind. Slowing down occurs when slow focusing is initiated, while speeding up occurs when rapid focusing takes place.

If you look again at figures 9.2 through 9.5, you will see how I depicted, through the use of blurs and focal points, the idea that any sequence represents either the growing control or the loss of control of an object. Habitual behavior occurs when the degree of uncertainty, both before and after an actual perception of reality, is minimized and hardly changes. In such a state, a person gains control over some aspect of her life. That control was illustrated by the amount of blurriness shown in the figures.

What I didn't tell you there was that this theory as shown in these figures also illustrates the degrees of certainty one has concerning who is in control. Moreover, it points to ancient wisdom (as I think the deceased spiritual teacher Krishnamurti used to mention years ago) suggesting that the observer and the observed are one. The blurred pictures show the spread of ego over different possible universes, as well as the uncertainty of position of an object in space. The greater the blur, the less the egoistic control. In other words, blurring also represents surrendering the ego. You can think of each little sphere in any blur of spheres as a separate ego in a parallel universe. The more parallel universes present, the greater the blur and the less presence of ego in any single universe.

Hence, defocusing actually means entering parallel universes, but with less grasp or control in any one universe. Focusing, on the other hand, not only provides certainty, it also provides a stable ego able to ascertain that certainty in a universe. Uncertainty not only indicates unknowing of a position of an object in spacetime, it also indicates a letting go of the ego by spreading it out over parallel universes. The greater the number of parallel universes, the greater the uncertainty, according to quantum physics, and the lesser the ability for an ego to gain control.

Although it may seem difficult to believe, I time travel everyday. I do so by simply gaining or surrendering control of my ego in holding onto or letting go of fixed ideas I have about myself and others. Letting go of my ego allows me to defocus and to move backward in time. Gaining control of my ego allows me to go forward in time. In this way, we can practice and naturally realize time travel in our everyday lives. It all depends on the intention of the observer.

Let me give you some examples. First, we need to emphasize that time is not just "out there," even though we observe objective processes as if they occur "in time/out there." Remember, time is created by the mind. As I mentioned at the start of this chapter, so are rate changes in its flow. The intervals of time we objectively mark as actual experiences appear as punctuations of consciousness. When many such punctuation (focal) points occur, we count them and mark them as "time's passing." In this manner, we acquire a "sense of time." If, say, a hundred such focal points occur that through habit we believe take a hundred seconds of objective time, we would afterward think that the same amount of time has passed, regardless of how much the clock-on-the-wall time has advanced.

For example, if I experience a hundred "second–units" in one objective second, as when I experience myself moving consciously and rapidly (as athletes typically experience in such sports as downhill skiing), time around me seems to stand still. In this case, I am experiencing a tightening of my ego—perhaps greater concern for my safety—which speeds time up. People who suffer automobile accidents often have this experience of "time-speeding." Notice that I refer here to the *sensation* I have of speeding, not to the objective amount of clock-on-the-wall time, which for me has slowed down. People who have "out-of-body" (OOB) experiences also feel a tightening of the ego's hold on the mind. As a result, they have the sensation of experiencing many subjective events while little objective time has actually passed. I realize that this statement may seem counter to what you may have thought takes place in an OOB experience; i.e., that in such a state, one is surrendering the ego. Actually, however, the very opposite occurs, in which the centers of focus shift from the whole body to the brain. Hence, the ego contracts even more, and the information received is even more illusory than in a normal state of consciousness.

On the other hand, when I experience a hundred "second-units" in slow meditation, time around me seems to speed up, and I experience "time-slowing" relative to the clock on the wall. I also experience time-slowing—which means I age less rapidly, even though the clock says I have aged normally—when I am creative and writing, as I am accustomed to do every day. Typically, I may engage in writing for a period of hours that actually appears to me to take only minutes. Artists often experience this "time-slowing" when engaged in creative activity. Here a surrendering of the ego occurs, and I seem to lose any sense of who it is that is doing the creating. People who act as channels know very well what this experience is all about. In a very real sense, to me creative activity not only takes me out of my ego, it allows my access to parallel universes where I can pick up lots of new information.

Hence, I am able to time travel at different rates into the future and into parallel universes depending on the degree of loosening or tightening of my ego.

But what about going back in time? According to quantum-physics principles, what we think of as the "real" past is also alterable and not fixed. This is certainly one of the strangest ideas to come out of quantum physics, yet we know from the uncertainty principle that the past cannot be absolutely pinned down and that consequently, one can change it by remembering it differently.

Let me briefly explain: Heisenberg's uncertainty principle concerns the observation of an object. It tells us that there are limits to the accuracy with which we can make concurrent observations. Take, for instance, an experiment in which either a particle's location in space or its momentum are to be observed. According to the uncertainty principle, these two possible observations are viewed as a pair of complementary observations. When one of these observations is made with great accuracy, the accuracy of the complementary measurement is reciprocally less determined. When position is well-measured, momentum is poorly determined, and vice versa. A similar tradeoff involves time intervals and energy changes in an object: When the energy involved in a process is determined as accurately as possible, the timing of the process will be undeterminable. This means we can't determine when the process takes place.

Consequently, in certain experiments it is possible actually to change our reconstruction of past sequences of events by changes we make in the present or future observations.[4] Such changes not only alter our memories, they also alter what we can legitimately reconstruct as the history of the events leading up to the present or future change. Does the "real" history change? While I may say it does according to my theory presented here, and others may say it doesn't because of their beliefs, how could we prove it either way? My point remains that only our mutually agreed memory—

our objective reconstruction of time—provides us with a picture of a frozen past. Hence, the question of just what the "real" past consists of remains a moot point. From my point of view, there is no such thing as an absolutely "real" history.

Consider the reexamination of your own thoughts. In accordance with the uncertainty principle, such an examination can alter your past recollection and the past as well. (Of course, what we mean by the *past* depends on more than one person's point of view.) What seems clear from this conjecture is that not only can a change in the future or the present alter your memory of the past, it can also alter your own participation in that newly remembered past. That's what physics tells me can happen, at least, in spite of its seeming science-fiction-like appearance.

Bear with me a moment longer. Remember that unfocusing appears as a loosening of the ego's hold in the universe. Consequently, in the parallel-universes way of thinking, this relaxation allows the mind access to other universes where different pasts could have taken place. As we may imagine, our memories play a much larger role in traveling back in time to parallel universes than they do in traveling forward to future ones. Time travel to past parallel universes may afford us the opportunity to alter former unfavorable situations in the direction of something more desirable for all those we encounter. To be sure, if they haven't made the trip with us, they will not change in spite of our efforts. But what we wish for others we see in the past will alter our memories of them, and certainly *we* will change.

However, I am saying more than would the theory of cognitive therapy, which posits that, while we cannot change the past, we can change our interpretation of it and thus its effect on us. If my speculation about time travel to parallel universes is correct, we can alter not only our memories of the past but even our participation in it as we reconnect a new past parallel universe with our present universe.

Changing the past, though, *can* lead to trouble if not done carefully. As I described in chapter 6, since a visit to the past involves going to a parallel universe, we will find duplicates of ourselves in it. For example, suppose that while we are "visiting" a past parallel universe we witness our past self involved in a car accident. Even though we hadn't suffered such an accident in the "real" historical past, when we return to the present, we now remember, not our accident-free history, but our accident. Far fetched? Well, suppose we are uncertain as to whether we (or was it our parallel self?) had the accident and ask another person who knew us back then. Suppose that person "remembers" that we did suffer an accident. Would we then be convinced? But wait a minute. Suppose we then go to our insurance company to see if an accident report was filed. The insurance company we "remember" we were with at the time tells us that indeed we did suffer an accident. Are we now convinced? Well, what has happened? As far as we now know, our "memory" of an accident-free universe has become only a fantasy, while the memory of the accident has become the "real" past. We have switched universes!

Does this mean that we can all simply change the past by wishing it to be so? Not at all. What happened in the above story would indeed be a shift in time between parallel universes. If it occurred, we would say we became confused and had a memory lapse. According to quantum physics, such a shift is possible. What's more, the history that develops in a shift from one universe to another will be logically consistent, and we will experience it to be the only universe there is.

Certainly, messing around in the past seems to be something we should avoid if we don't know what the consequences are likely to be. However, if events in our present can alter history, then the past that you remember is continually changing, whether or not we actually realize it. The past as recorded in history books, for example, will always be subject to the writer's recall, imagination, and rationalization. What is written as history is not the "real" past, which even in the memories of those that experienced it is subject to change. I think someone once said: "History is written by the winners." Thus those who write history may very well be altering it to make themselves appear more favorable. Of course, we may say that the historian has simply reinterpreted history, but that the "real" history, which may be unknown, has not changed. Although that statement certainly makes common sense, it is ultimately unprovable because, according to quantum physics, there is no such thing as the "real" history: The past cannot be pinned down, even though we commonly refer to it as if it was.

In fact, different versions of the past may be the cause of many troubles in our world today. Can you think of any current world conflict that hasn't resulted because of diverse accountings of the participants' "common" past? I cannot. Perhaps the realization that the past will always be changeable, and that even our memory of it can change, will give us a greater ability to be tolerant of other views.

I know it is nearly unthinkable, and perhaps seems even ridiculous, to entertain such notions, but the past is not fixed in spite of our present memories. Each time a switch takes place reconnecting a new past parallel universe with the present universe, history changes as we remember it, and we emerge in the new parallel universe with memories now consistent with it. Certainly such shifts are indeed usually very tiny, but parallel universes are closer to us than we might imagine. The key to accepting this concept is in releasing our hold on the idea that there is only one past back there in time.

Thus, our own personal history of our growing-up years, our past friends, where we lived, and so on, appears quite fixed in our minds. If it were to change even slightly, through a shift to a parallel universe, the new memory would be just as fixed in our "newly" made-up mind, and we would have no memory of the way it was. These "blind" shifts would not be remembered at all. We may thus be holding at this very moment alternate memories of our parallel self, all the while thinking that we have held these memories all our life. Indeed this is strange business, but quantum physics says it's so.[5]

However, for big changes to be made in the past involving parallel universes, my research suggests that many people would be needed. It works like a hologram: the more area of the hologram being illumined, the stronger the "signal," and the greater and more real becomes the image. Smaller changes—individual changes—can be accomplished individually or in a small group.

For example, think of the common practice of gathering in worship. If (hopefully) egos are surrendered, each participant has the opportunity to use his or her mind to time travel. Many people in worship tend to remember past transgressions. But is this actually time travel? For example, in the Jewish religion, the day called Yom Kippur is a day of remembrance and forgiveness. If what I am proposing is accurate, during their practice in such a gathering the worshipers can travel back in time to their ancient roots. Regardless, it would help the world if such a practice was more in the way of forgiving than simply of remembering, particularly if those holding that memory were continuing to hold a grudge. Certainly, participants in worship find benefit in their practice if they *do* surrender their egos. If they forgive, they will change the world. And if I am correct, they will also have time traveled.

On the other hand, Iyengar hatha yoga practitioners tell me that body memory often works as a detriment to growth and change. Your memories act as the chains felt by the prisoners of Plato's cave. You may be holding a particular posture incorrectly for long periods of time because your body holds a memory that prevents you from a fuller extension. Using time-travel techniques—loosening the egoistic concerns (such as "I won't do this because I will look foolish" or "I'm afraid")—could help you speed up the learning process just by changing your past beliefs and forgiving yourself. Simply stop putting yourself down for past errors of judgment and realize that a particular memory may have served you well in your past but no longer needs to be a chain holding you there.

In this light, a group that attempts to practice worship without really letting go of the ego will not time travel and as a consequence will not receive any benefit thereby, simply because no one in the group will have succeeded in the voyage. Fear and unforgivingness only result in a replay of the old movie we cannot let go of.

Shamans often take the tribe or a group of individuals within the tribe on a time-travel journey, usually back in time but also forward, to help the tribe cope with coming environmental changes. If our own community spiritual leaders were to take on a similar shamanic role, I believe that their communities would be helped in the attempts to resolve past-generated grievances. For example, those that practice Transcendental Meditation in large groups have found that such practice has an effect on the community at large by reducing crime as well as slowing aging in individuals who meditate.[6] While the Maharishi University of Management does not claim these benefits are due to time travel, they have established that Transcendental Meditation has a very profound basis in quantum physics.

The key ingredient here is realization of the sacred aspect of time travel: Give up your ego, enter the sacred timeless realm, and forgive. Such is the path to true freedom.

As we become more aware of just what time travel really is, and we learn that the old model of linear time inevitably marching on no longer holds in our now post-Einsteinian and post-quantum physics world, the question of how and why time travel improves our lives will become more and more evident. The crucial step is in understanding our own egos, why we have created them, and how we can control and let go of them. Not learning how and when to tighten and loosen our egos has a price; it robs us from our happiness.

Some of you may be getting the idea that I am saying the ego is a bad thing and should be dropped. Not at all. Ego has a very important place in the universe; otherwise, it would have never arisen. Ego provides a deep sense of self and other and an awareness of life and death as well as a basis for experiencing the material world. It enables all of the wonderful individual expressions of life we find in other people different from ourselves. Because it anchors us in time and space, it provides us with the opportunity for a deep appreciation of the world. It also enables despair and longing to arise when our needs are not met.

The longing for positive change or spiritual change is really a waking up from the grand illusion brought on by our belief in linear, ever-increasing time. From a timeless or spiritual base of understanding, nothing really changes. There is a flow without time's presence, so that each of us realizes we aren't really separate beings limited by space and time, but a One-Being, continuous and eternal. This realization came to me from my understanding of the physics of our universe as much as it came from my own spiritual realization.

We can all learn to reverse time, certainly for short intervals, by letting go of past fixations that tend to make us automatically predict the future. These "cause/effect" relations project the normal flow of time we tend to objectify and hold as the only way that time can go. I hasten to add that some past fixations, such as good motor-vehicle driving habits, should not be let go of! We need them to be able to predict the other driver's behavior, for example. Usually this keeps us safe. But some fixations we have learned through habitual behavior do not help us at all, and by letting them go we are actually turning our internal clocks backward. For example, if you have a bad habit that you continue to support, think of it as something that ages you unnecessarily. Breaking that habit will not only stop unnecessary aging but will reverse your internal clock, making you younger.

Remember that our past viewpoints will not necessarily be the ones we have currently. Through the practice of mind yoga, we will be changing who and what we think we are; our choices will change when we change our self-concepts. There are no limits to what we can see in the past or future.

The meaning of our own deaths will change rather radically when we grasp what time travel implies about the self and the soul. From my quantum-physics point of view, conscious life consists of patterns of focused and unfocused activity. As we see, these patterns give rise to the ego or body-mind, which arose evolutionarily as a mechanism for survival. Death, being the release of that survival mechanism, returns the observer or Mind of God to the timeless realm wherein focusing on objects needed for survival, such as the body and its immune responses, no longer occurs or is required. Time travel, according to mind yoga, temporarily relieves the Mind of God from the activity of focusing and enables God to expand across time. Thus time travel defines our relationship with God.

Is reincarnation a viable idea in light of time traveling? If reincarnation means that each of us returns intact with the same ego, I think it is unlikely. My view posits that consciousness or mind itself is not unique to individual or even human bodies, but is universal and nonlocal. Individual consciousness is a necessary distortion of the Mind of God needed for creature survival in much the same way that we need our projection mechanisms in order to make decisions about the future, or even to see that an object is "out there." We know that mental projection distorts reality or creates an illusion of it, yet we are still dependent on it for our survival. From the time-traveling perspective offered here, we see that ego surrender allows new possibilities to arise. What may seem surprising is that as far as physics is concerned we can reincarnate in the past as well as the future. In this light, consider that Buddhists believe in a form of reincarnation different from individual consciousness survival. They tend to see reincarnation as a process, which propagates tendencies through time rather than actual personalities. From their vision of a timeless-deathless realm in which no one is born and no one ever dies, I suggest that reincarnation in the past is just as likely as reincarnation in the future.

We have seen that time travel is intimately tied to modern physics, and that temporal concepts of physics coincide with sacred vision. Time travel involves a change in ego structure—meaning a change in an individual's mind as it relates to spacetime. By loosening the ego's boundaries, one experiences oneself at several times simultaneously. This is the natural arena of subspacetime—the world of the Soul and of God. In ancient times, we might have called this "heaven."

Realize that we all are participating in an awe-inspiring journey given to us by our temporarily forgetting who we really are—the Mind of God. That concept may sound pretentious, but it has become and remains my faith.

CHAPTER ELEVEN

Summary *and* Conclusion

I've been on a calendar, but never on time.
—Marilyn Monroe

Whoever controls the past controls the future.
Whoever controls the present controls the past.
—George Orwell

Let's review where we have been on this journey into the ways of time travel. In the introduction, we explored the meaning of time travel and the roles played by memory and attention. We considered several different forms of time travel, and we reviewed some of the basic principles of physics involved in our notion of time. We also went into some of the perplexing time-travel paradoxes that arise simply because we aren't used to thinking beyond our linear motion of time. I offered the amusing speculation, based on Lewis Carroll's delightful *Alice in Wonderland,* that our memories might work better if we could recall the future as well as we recall the past. This idea began to make more sense as we saw how our experiences in time are created from a sub-spacetime realm.

In chapter 1, we saw why the teachings of yoga are pertinent to time travel. We looked through the history of time travel as seen by ancient Indian people, and we reviewed how ancient Indian yogis thought about time, seeing it as a great god. These yogis described ways to cheat the Time god so that he would not gain hold over their lives. I related the teachings of the Bhagavad-Gita to our Western view of time and discussed the conflicts felt by Arjuna when he is confronted with having to go into battle against loved ones. Krishna tells Arjuna that He is Time and that from His point of view the battle is already over and the outcomes determined. Krishna appears to Arjuna in many forms, as if Krishna was showing himself as a *possibility*-wave.

Krishna explains the reason we are caught in time. He, Krishna, created desire, and we express this desire in terms of the physical realm and in so doing become captured by it. While desire for a material universe ultimately "creates" the universe, the fundamental desire in individuals, whether they remember it or not, is the yearning to be Krishna. So Krishna gives every sentient form the ability to focus and defocus *possibility*-waves and thus the means for evolution, happiness, and sadness, which arise from the probability-curves we each generate in our attempts to master the physical world.

Let's consider the nature of desire a little farther. The Gita talks about action and introduces the idea of an "actionless" action.[1] The difference is subtle. Normal action can do great good or evil and, since it is always accompanied by the ego-self, is concerned with the "fruits" of the action. In this case, "fruit" refers to the transformation from *possibility*-wave to probability-curve—a process that we saw depended on the pattern of collapses or focuses as I called them. On the other hand, actionless action works at the level of *possibility*-waves, and appears as selfless action when its consequences finally manifest. Hence, actionless action has an effect over a greater time period, even though its immediate consequences may appear at local times and places.

In chapter 2, in order to grasp the spiritual quality of the timeless spaceless realm, we compared ordinary scientific space and time with spiritual, or sacred, space and time. In brief, to make use of a geometric metaphor, scientific spacetime is linear while sacred spacetime is circular. We saw how the cyclical time of sacred tribal peoples and Indian philosophy touches base with the linear time of Western culture. I introduced the similarities and differences between line-time and cycle-time and specifically discussed how Australian aboriginal peoples made this connection.

Indian philosophy regards the timeless realm as more real than the manifested realm confined by time and space and says the task of conscious beings is to discover that timelessness and give up any hold one has on the time-space-matter universe. It reassures us that in spite of our common experiences, needs for survival, and fear of death, our basic nature is not subject to life and death. However, we cannot realize this truth so long as we remain consumed with life's ever-present dualities. The key insight comes when we are able to detach ourselves from these everyday concerns. Difficult as this task may be, it is the same difficulty posed by time travel itself, for freeing oneself from life's dualities also constitutes a necessary step in time travel. In brief, the key to traveling through time is to free yourself from your everyday concerns.

Chapter 3 presented the basic reasons why physicists now take time travel seriously. We explored Einstein's general and special theories of relativity and how they have changed our commonsense views about space and time. We saw, in a somewhat historic sense, just how important, although normally invisible, a role time plays in our everyday lives and how it provides the foundation of all sciences such as geology, biology, astronomy, and physics.

In chapter 4, we inquired more deeply into how we think about time and space. We examined different metaphors for time travel, and I explained how they reflect time but don't really describe it correctly. Then we considered how certain adepts of India mastered time travel, specifically looking at the life of one adept, Sri Ramana Maharshi. We began to see the role played by the self or ego trapped in time and space. We saw how the actions of the ego actually change our sense of time. We discovered that the seemingly ever-forward flowing movement of time is changed in the brain and began to see that consciousness and time are therefore intimately related and, indeed, may be the same thing. We looked at industrial machines and how the need to ease work led to the thermodynamic arrow of time, which provides a direction or sense of a normal "flow" of time. Finally, we compared thermodynamic, industrial linear time with the sacred hoop time of ancient aborigines, adding to the concepts introduced in chapter 2.

In chapter 5, we took a *tour de force* through space, time, energy, and time travel as seen by modern physics, with the aim of making the idea of time travel clearer than before. Reviewing the delightful time-travel book *The Time Machine* by H. G. Wells, we saw that Wells's vision included elements of Einstein's ideas, even though they were published ten years earlier. This chapter dealt more directly with relativistic physics, emphasizing its rather amazing and entertaining range. We looked, for instance, at the connection between gravity and time and how gravity affects time by slowing it down. We then explored gravity in terms of Einstein's equations, which predict black holes, and how in certain circumstances black holes become wormholes, which could be used as shortcut tunnels through space or as time machines. We saw several examples in detail.

We saw how although the general theory of relativity predicts time travel, there are still problems to consider, specifically, the paradoxes that time travel introduces. In chapter 6, we looked into the two major paradoxes: the knowledge paradox and the grandfather paradox. The knowledge paradox has to do with what happens when future knowledge is brought back in time to the past. If a future person gives his younger self certain knowledge, and the younger self retains that knowledge and, upon growing up, uses it in the future and also hands it back in time to his younger self, this cycle can repeat *ad nauseam.* Who, then, created or discovered this knowledge originally? The grandfather paradox is concerned with what happens if a person goes back in time and arranges things such that her parents do not meet, so she is never born. How can she exist in the future and thereby travel back in time?

Next we explored the question in chapter 6: Can we get around these paradoxes, or are they fatal to the possibility of time travel as a reality? We discussed the bizarre notion that we can indeed get around these paradoxes via the parallel universes envisioned by quantum physics. Hence it appears that not only is it possible to time travel to both the past and future, the process is woven deeply into the fabric of physics.

In chapter 7, we looked at how we might engage in time travel using various devices and technologies currently under theoretical development and planned for the future. My purpose in presenting these was to make clear how quantum physics opens the door to time travel, and, since quantum physics deals with probability in an essential manner, how time travel and probability are deeply related. We looked at ways we could use technology to build a time machine, specifically, a device envisioned by physicists Aharonov and Vaidman that would shift time inside of a massively heavy, hollowed-out sphere. The setup used would use a quantum computer—a device that uses qubits, or number-possibilities, rather than numbers. By putting the quantum computer into a special state, the *possibility*-wave of the person inside the device could be shifted in time either forward or backward, thus producing counterfactual realities—a future that would have been improbable or a past that had not taken place.

Chapter 8 was a key chapter. We saw how possibility, probability, and time are related. We looked at what *possibility*-waves accomplish and why they are necessary. We discovered that they operate in a sub-spacetime reality and connect to spacetime reality through a simple mathematical operation called squaring, which changes these waves into probability-curves. The ability to square affects everything that manifests, including those who carry out the operations. I showed how the results of the squaring process depend critically on *superposition*—the overlapping of two or more *possibility*-waves. These waves can add together producing a new *possibility*-wave. When the new wave is squared, it produces a very different probability-curve from the one produced by squaring the *possibility*-waves first, before making up the sum. Consequently, the squared new probability-curve gives rise to new behavior.

I introduced several examples based on gambling house practices to clarify just how this process of squaring and then adding, or adding and then squaring, works. We looked carefully into the question of when squaring should occur, either before we add *possibility*-waves or after. The answer gave insights into habit formation and thus ways to alter our behavior by changing the structure of events in time. We saw that waves going backward in time mix with or modulate waves coming forward in time. The mix appears as a probability-curve. Hence, we learned that probability at any instance, although normally statistically independent from the past or future, does in fact depend on the past or on future expectation. Perhaps this explains why people often feel they are either luckier or unluckier than others.

In chapter 9, we looked at how the notion of time arises and investigated mind yoga as a means for time travel. We saw how habit and probability are deeply connected and how consciousness itself comes into the mix, acting as a time machine—either speeding the flow of time or slowing it down. We saw how freeing the mind from the body—paradoxically, by engaging it more consciously with the body—enables the person's self-concept to dissolve. The "I" or ego-self alters *possibility*-waves, producing probabilities for real events in the mind. The sequential order of these probabilities produces time order. Dissolution of self-boundaries enables the experience of time travel. As the time traveler/yogi's body awareness changes, a quantum-physics state of consciousness arises, projecting the mind into a future or past time state. We examined the function of sub-spacetime in this process—an imaginal realm in modern physics similar to Plato's world of ideals. It appears that the existence of this realm provides a "solid" ground of being for the physical world we are able to measure and sense.

After this brief examination of the role of sub-spacetime, I suggested that its existence has a connection to spirituality. The reason for this suggestion is the following: Modern physics, particularly quantum physics and the general theory of relativity, tell us that time is not as our classical physics heritage would imply. In light of what we are learning now, time travel becomes more than a possibility; it becomes a necessity. An old adage in physics says that whatever is not forbidden becomes compulsory. In the last decade or so, there has been a big change in the scientific attitude toward time travel. Originally, the burden was on physicists to prove that time travel was possible. Now the burden of proof has shifted to proving there is a law forbidding it. It now seems that time travel may be an essential requirement in physics' expanding menu of remarkable phenomena.

With this change, our understanding of time is shifting yet again, the importance of consciousness as an element in physics is becoming apparent, and the link between time and consciousness has been forged. The seat of consciousness—the soul or essential self—now appears to be directly involved with time, possibly with its very emergence as something we think we can objectify. The role of consciousness has been presented variously by several physicists and physics/consciousness pioneers, such as Henry Stapp, Amit Goswami, and Roger Penrose.[2] Consciousness acts or has an effect on physical matter by making choices that then become manifest. It now appears that such an action cannot simply take place mechanically. Implied now is a "chooser," or subject, who affects the brain and nervous system. Some physicists, such as Stapp, believe that this chooser arises in the brain through past conditioning. Penrose believes the action of choosing takes place nonalgorithmically—that is, not through the action of any mathematical formula or any computer-like process. I suggest that this chooser/observer does not exist in spacetime and is not material, which suggests that it is a spiritual essence or being residing outside of spacetime.

The choices that the chooser chooses show as *possibility*-waves existing in a sub-spacetime realm. They move about in this realm oblivious of both space and time, existing in several places simultaneously and even in the past and future at the same moment. Attempting to describe how things can exist side-by-side in the future, the past, and in several places at once seems to push language to its limits.

When consciousness acts, *possibility*-waves traveling backward through time modulate waves traveling forward through time. This modulation results in the squaring process that yields a probability-curve that makes sense in our physical world. The probability-curve provides us with opportunities to control our lives, for they enable us to develop habits of behavior and expectations for future success in any endeavors we pursue. Without this squaring action, the world would remain in a timeless, spaceless state of ever-changing possibilities, with nothing ever manifesting at all. Strange as it may seem, there would be no consciousness of any object appearing anywhere or anytime. Indeed, there would be no time nor space wherein anything physical could appear.

The squaring procedure results in a pattern of *possibility*-waves moving across periods of time, thereby producing the past and future as well as the present. This pattern, stretching over time through its multiple reflections, gives rise to self-consciousness and creates within spacetime the ego structure, which allows individual evolutionary behavior, survival, and spiritual awareness. Hence the ego structure or self-concept exists as a pattern in spacetime. Yoga, through its body poses, enables individuals to change this patterning. Mind yoga offers a similar change, providing the ability to move in and out of the timeless realm—that is, to time travel.

In chapter 10, after reflecting on how physicists today appear to need this sub-spacetime world in order to explain the physical world, we explored the spiritual dimension of time travel. I presented a view of the sub-spacetime world as both a spiritual and physical necessity. I believe that just as time travel is now emerging as a mainstream concept in our world, the sub-spacetime world will also emerge as mainstream in the near future.

Because we are embodied and thus limited by space and time, we sentient beings have lost the experience (if it can be called that) of residing in the timeless-spaceless realm. Even bringing up the concept boggles the mind and raises hackles of suspicion and skepticism. For centuries, spiritual practices, including meditation and various forms of ritual, have been utilized to break free of these limitations—to join us with our predecessors who have so acted in the past and to link us with those in the future who will continue the tradition. Spiritual teachers, realizing this difficulty of seeing beyond the everyday reality of space and time, ask us to practice techniques known experientially to assist one in this endeavor. Paradoxically, even though I call it an "endeavor," the techniques are really designed to break out of the endeavoring that has become habitual to us.

I once heard the Dalai Lama explain why he spends so much time daily in meditation. I was surprised that he framed his answer in terms of what happens to a person after death. He, like many other spiritual teachers, explained that the mind continues after death; however, it is easily distracted. To prepare for this transition, the Tibetan Buddhist trains his or her mind to understand the deepest levels of existence. The Dalai Lama's daily practice consists of as many as seven separate periods of meditation:

> My daily practice is preparation for death through meditation—to make a separation of the body and consciousness. Unless you reach the deepest sub-consciousness, you cannot separate this body and mind. When you reach this deepest state, then they can separate. Then I go deeper, deeper, deeper, to the deepest which is the clear light. Sometimes I joke: In 24 hours I experience death and rebirth seven times. Then when the actual death happens, this practice becomes very useful.[3]

Like death and deep meditation, the mind yoga form of time travel frees one from space-time limitation for the periods when the time travel is projected. In other words, while the rest of us hold on to our timekeeping in the usual manner on earth—and the mu mesons hold on to their timekeeping while they speed through the universe possibly at so near lightspeed that they may ride for years of our time but less than a microsecond of theirs, and while distant parts of our universe rush away from us at so near lightspeed that in an hour for us millions of years pass for them—we in meditation are free of time altogether.

This respite from the space-time continuum allows us to move backward in time and revisit past "errors." It enables us to forgive ourselves and others because we see that the other is nothing more than the self. It offers a fresh, new life-affirming quality to everyday existence. Perhaps most importantly, while engaged in this kind of travel you slow your body's aging processes, simply because the mind has relaxed its concern with creating an objective world arising from probability-curves. By using meditation and yoga techniques, though it may not seem to be so, you are freeing yourself from time. With practice, as Ramana Maharshi showed us, you will be able to "see" into the past or future as well as beyond your own immediate spatial region. Ultimately, as Maharshi indicated, the key to time travel seems to be in surrendering egoistic patterns. Traveling into the future may, in this way, hold the possibility of a truly more peaceful world.

Finally, let's take one more look at what Patanjali said about time. In the fourth chapter of the Yoga Sutra, he tells us that past and future exist as different condition of the eternal now.

> Pure soul awareness is the true Yogi, which is changeless and non-moving, its form having accomplished its own intelligence, assumes the identity of knowing. Time—the sequence of the modifications of the ego mind—likewise terminates, giving place to the Eternal Now. Consequently total liberation becomes possible when the three qualities of matter (light, inertia, and vibration) no longer exercise any hold over the Yogi as well as having discharged the four-fold aims (duties to self, family, society and country). Once established in one's own true nature, the power of pure soul awareness, there is nothing left to be done.[4]

APPENDIX

CLASSICAL *and* QUANTUM COMPUTERS

In case you are interested in how modern and future computers work, I have included these "bits" on *bit*-classical computers and *qubit*-quantum computers so that you can clearly see how they differ.

ORDINARY "BIT"-CLASSICAL COMPUTERS

In our modern world, lots of things are made to be turned on or off. This is particularly true for computers. Like the instruction you leave in a note for your children to turn off the lights when they go to bed, a computer program consists of a similar set of instructions called a "code." It contains a string of on and off symbols providing instructions telling some particular memory register just where to turn on or off and when to do it. Any number, such as a house number or address, can be written as a string of on and off symbols. Hence a computer's code string or instruction typically seeks an address, just like you would look for a house number. Then it leaves a note with the instructions: off—on—on—off, and so on.

An instruction of either on or off is called a "binary code." Computers produce numbers that are combinations of what are called "bits"—actually just 0s and 1s that are no more than simple "offs" and "ons," with 0 meaning "turn off" and 1 meaning "turn on" or vice-versa, depending on the manufacturer of the computer and subtle code changes such as ciphers and encryption techniques. Strings of bits don't make a necklace, but they do make a very specific instruction in binary code. Modern computers, of course, have the means to translate binary code into decimal numbers, or words of English, German, Japanese, and so on, so that what you see on your screen or printed page is recognizable script.

1-1-0-0-1-0-0-0

Figure a.1. *A binary-code instruction.*

But before you see a number on a screen or a printed page, the computer must follow a program. It must undergo a very rapid series of "additions" and shifts to the left or right of a given bit in a memory register before it stops. For example, a register of an eight-bit computer might look something like figure a.1.

This particular string of 0s and 1s contains eight bits and represents the number 200 written in binary arithmetic. Strange-looking as it is, it represents an instructional code to the computer. That instruction begins with the bit appearing on the far right—in this case "0"—and continues along the string one bit at a time until it reaches the last bit on the far left.

Each position in the register represents a specific number to be added (or not added if a 0 appears at the position) to a total number in the following sequence: 1, 2, 4, 8, 16, 32, 64, and 128. You may recognize this sequence as the numbers generated from increasing powers of two. The starting number is the number 2 raised to the "zeroeth" power. (Mathematicians observe the convention that any number raised to the zeroeth power equals 1. Thus $25^0 = 1$, $342^0 = 1$, and so on.) If a number 1 had appeared at the corresponding position in the register, it would mean that

2 raised to the "zeroeth" power must be added to the total. Since the number 0 appears there, it means not to do this. Continuing along, the next register's position has the number 1, and so on, to the last register position with the number 1. Here is what that above code says:

1. Start with the number 1 (2 raised to the 0th power) and multiply by 0 (the number appearing at this position), then shift left.
2. Add the number 2 raised to the 1st power (2) and multiply by 0 (the number appearing at this position), then shift left.
3. Add the number 2 raised to the 2nd power (4) and multiply by 0 (the number appearing at this position), then shift left.
4. Add the number 2 raised to the 3rd power (8) and multiply by 1 (the number appearing at this position), then shift left.

And so on. By doing this you construct the number 200 as the sum of numbers consisting of 2s raised to various powers: $2^3 + 2^6 + 2^7 = 8 + 64 + 128$.

In the instructions just above, I said multiply by 0 or 1. Actually, ordinary computers do not really multiply—they only add and shift. Hence "multiply by 0" simply means add nothing to the sum, and "multiply by 1" means add 2 to the power indicated by the register number's position from the right to the sum in question.

The computer carries out its instructions without really making any attempt to insert meanings for the numbers such as I have laid out above. It just adds according to the usual rule of binary addition. That rule is $0 + 0 = 0$, $0 + 1 = 1$, $1 + 1 = 0$ (shift left, add 1). Actually this is very similar to what you do when you add two numbers together, such as $6 + 5 = 11$. In ordinary math, we use a decimal system, so our rule would be $6 + 5 = 1$ (shift left, add 1).

In summary: A number in an ordinary computer consists of a string of bits—0s and 1s—that composes an instruction sequence telling a particular memory register how to change itself, or not, and in what order to do so.

Extraordinary "Qubit"-Quantum Computers

Quantum computers resemble ordinary computers, yet they operate in a very different manner. Their memory registers are schizophrenic—they don't remember bits at all, but they do remember what a bit could possibly become. When we deal with possibilities for bits, we enter the quantum world where these *possibility*-bits are called *qubits*—or quantum bits of information.

How do a qubit and a bit differ? First, let's picture a bit using a two-position arrow. Since bits are simply on-and-off devices, they can point in only one of two possible directions. Figure a.2 represents a space for a bit with its two possibilities. Think of the two axes as arrow directions. The arrow in this two-possibility space can point either to the right or up, as shown in figure a.3.

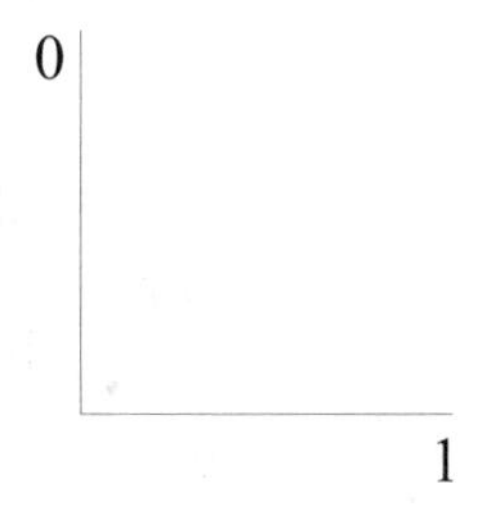

Figure a.2. *The bit space.*

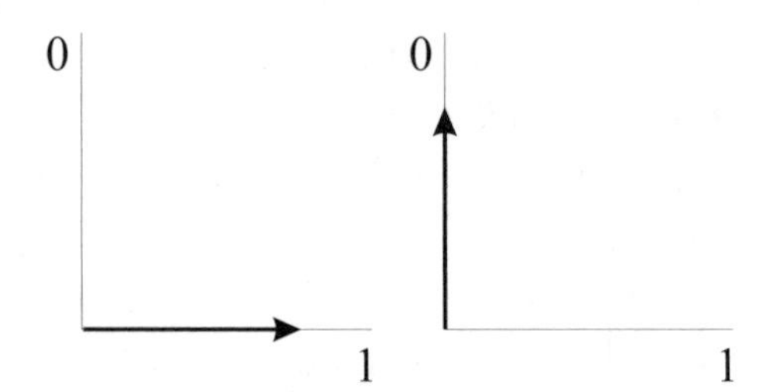

Figure a.3. *The bit space with two bits, 1 and 0.*

When the bit registers "1," the arrow points to the right, and when it registers "0," it points up. In figure a.3 we see a two-bit register showing the number 2. In principle, there is no limit to how many bits we can use in string. For example, an eight-bit register showing the number 200 is pictured in figure a.4. This is the same number as shown in figure a.1, but using arrows instead of 0s and 1s to show the values of the bits.

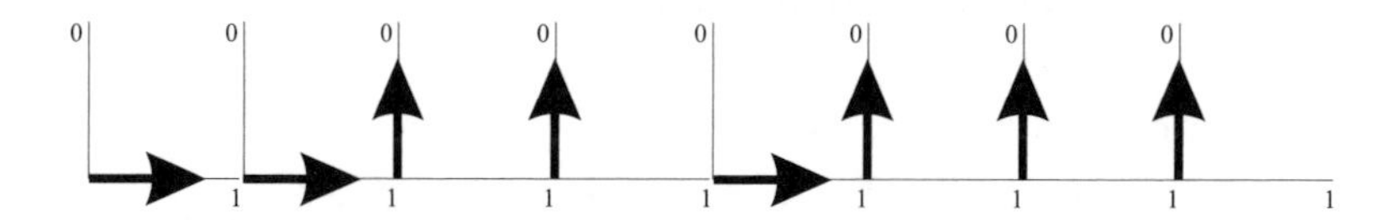

Figure a.4. *The eight-bit space with the number 200.*

Suppose we simplify to a string of just four bits. Figure a.5 uses four arrows to show the number 5. Remembering that 2 to the power 0 equals 1, we simply add the results by moving along from one register to the next. So we reach the number 5 by adding $0 \times 2^3 + 1 \times 2^2 + 0 \times 2^1 + 1 \times 2^0 = 0 + 4 + 0 + 1$. Following a similar procedure, we can construct from the sixteen different arrow positions any number between 0 and 15.

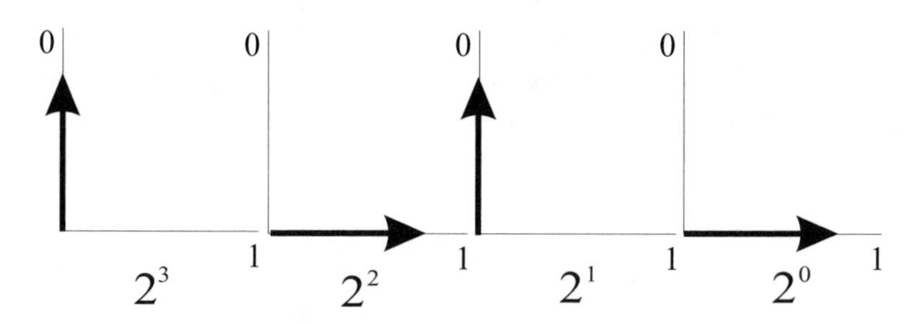

Figure a.5. *The number 5 in a simpler four-bit space.*

What about qubits? Qubits enter the picture when we consider what would happen if the arrows were not pointing completely off or on, but somewhere in between. Quantum physics is the science of "in betweens," not-quite-this-or-that, and possibilities. It lets us deal with arrows that are pointing neither up nor to the right.

Qubits are the "maybe's" of a quantum computer. They are to a quantum computer as bits are to an ordinary computer. A quantum computer allows users to play with arrows that never get completely thrown one way or the other while the computer computes. In fact, a computation is nothing more that a program that tells these arrows how to change their positions from one in-between direction to another.

Even though quantum computers work this way, "maybe's" are never seen in the "real" world—all we see are the realities of 0 or 1. Hence—and this is where the quantum world enters the real world—when we attempt to *find out* in which direction the arrow points, we usually find that it is pointing either to the right or up, as shown in figure a.3.

For this reason, quantum computers work best without us looking in on them and when their qubits move about in "maybe" positions. Since there are an infinite number of maybe's possible, the number of computational possibilities for quantum computers becomes well beyond the scope of any classical computer.

In figure a.6, we see a typical qubit represented with its arrow pointing halfway between a 0 and a 1. In figure a.7, we see a four-qubit array showing "something." Whatever it's showing, it's not a number. If we attempted to observe this register, the arrows would snap to either the 0 or the 1 position.

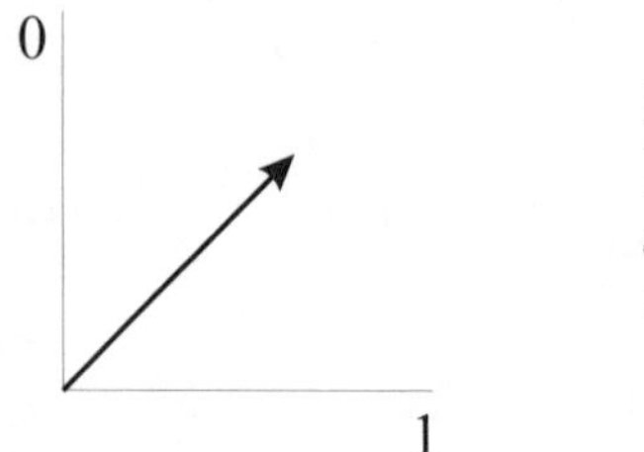

Figure a.6.
The bit space with a qubit—in a maybe 0 or a 1 position.

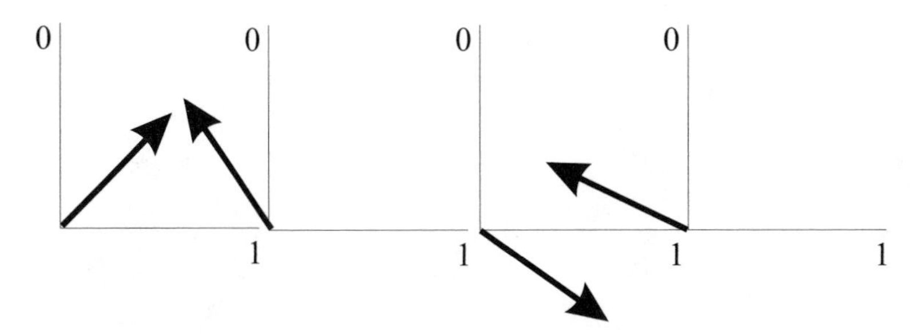

Figure a.7. *Not a number, but a possibility in a four-qubit space.*

There is a little more to the story of these remarkable devices, having to do with how we add qubits. It turns out that a special kind of addition involving superpositions of qubits is needed in a time machine. It is not always easy to picture such superpositions, even though the tiny qubit-registers in a quantum computer handle them very well. Figure a.8 shows one that can be displayed correctly. The superposition of a single qubit that is simultaneously a "1" and a "0" produces a "maybe" with its arrow pointing at a 45-degree angle.

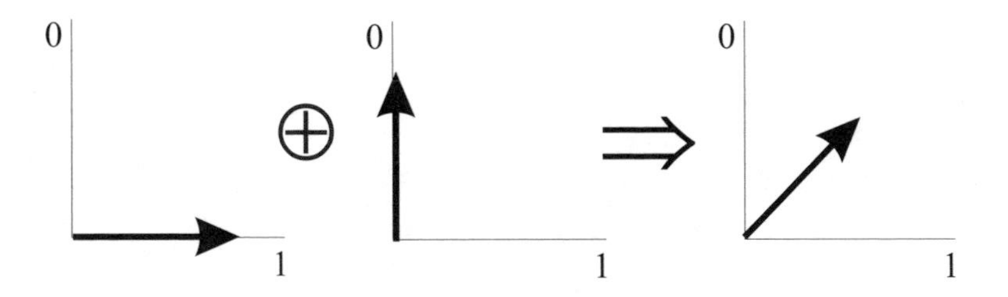

Figure a.8. *A single qubit superposition.*

Note my language here. That's right. This may look like I've added two distinct qubits, but I haven't. The computer would add these results in a single qubit register. An ordinary bit-register must contain either a 0 or a 1, but a qubit register can hold a maybe. And to get a time machine to work, we need a lot of maybes to come together.

NOTES

Introduction

1. Wilder Penfield, *The Mysteries of the Mind* (Princeton, NJ: Princeton University Press, 1975), 34.
2. B. Rossi and D. B. Hall, "Time Dilation—An Experiment with Mu-Mesons," *Physical Review* 59 (1941): 223.
3. Just what these remnants are depends on each muon's electrical sign. If it is positively charged, it leaves behind two kinds of neutrinos and a particle of antimatter called a positron. If it is negative, it leaves behind the same two kinds of neutrinos plus an ordinary electron. Either way, the decay event leaves a mark in a counter.
4. A microsecond is one millionth of a second. Typically, the length of time subatomic particles live before decaying is measurable on a scale of microseconds, but some particles live much longer. For example, a particle called a neutron (neutrons are found inside the nucleus of nearly every kind of atom) lives about fourteen minutes before it decays. Physicists measure a particle's lifetime in terms of its half-life.

   The half-life of any object is the amount of time it takes for 50 percent of a group of those objects to decay, that is, to break down and disappear. For instance, muons have a half-life of only 1.5 microseconds. That means that if you could gather one hundred muons and put them on a plate and watch them, after 1.5 microseconds half of them would be gone. After another 1.5 microseconds, half of those remaining fifty would be gone. After still another 1.5 microseconds, half of the remaining twenty-five would be gone, and so on.

   This same principle is the basis for the archeologists' tool known as carbon dating. A special isotope of carbon known as carbon-14 decays with a half-life of about five thousand years. All living bone has a certain percentage of carbon-14 that remains about the same so long as the creature is living. When the creature dies, the carbon-14 begins to decay, hence that percentage decreases. By measuring the level of carbon-14 in the bones of an animal—one that lived no more than five thousand years ago—archeologists can determine when the creature lived.

CHAPTER 1

1. Georg Feuerstein, *Tantra: The Path of Ecstasy* (Boston and London: Shambhala, 1988), 32.
2. Ibid.
3. Ibid., 378.
4. Eknath Easwaran, trans., *The Bhagavad Gita* (Tomalas, CA: Nilgiri Press, 1985).
5. Ibid. The revelation is in chapter 11—two-thirds of the way through the eighteen-chapter text.
6. See http://www.bhagavad-gita.us/bhagavad-gita-11-4.htm. Also see A. C. Bhaktivedanta Swami Prabhupada, *Bhagavad–Gita as It Is,* chapter 11, verse 7, in *The Library of Vedic Culture*, an interactive CD (Los Angeles: The Bhaktivedanta Book Trust, 2003).

CHAPTER 2

1. From "Time in the Indian Tradition," a lecture by Prof. E. C. G. Sudarshan, Department of Physics, University of Texas, Austin. http://www.here-now4u.de/eng/time_in_the_indian_tradition.htm.
2. Ibid.
3. See chapter 9, which describes the Dreamtime of aboriginal peoples of Australia: Fred Alan Wolf, *The Dreaming Universe: A Mind-Expanding Journey into the Realm Where Psyche and Physics Meet* (New York: Simon and Schuster, 1994; reprint: New York: Touchstone, 1995).

4. W. H. Stanner, *White Man Got No Dreaming: Essays 1938–1973* (Canberra: Australian National University Press, 1979).
5. B. Spencer and F. J. Gillen, *The Native Tribes of Central Australia* (New York: Dover, 1968 [first published in 1899]).
6. See Colin Dean, *The Australian Aboriginal Dreamtime: An Account of Its History, Cosmogenesis, Cosmology, and Ontology* (B.S. thesis, Deakin University, 1990; available from the Australian Institute of Aboriginal Studies, Canberra).
7. W. Love, "Was the Dream-Time Ever a Real-Time?" *Anthropological Society of Queensland Newsletter* 196 (1989): 1–14. Available from Australian Institute of Aboriginal Studies, Canberra.
8. Ebenezer Ademola Adejumo, *The Concept of Time in Yoruba, Australian Aboriginal and Western Cultures, Especially as It Is Manifested in the Visual Arts* (M.S. thesis, Flinders University of South Australia, 1976).

Chapter 3

1. Saint Augustine, *Confessions,* bk. 11, chap. 14. A complete copy of Saint Augustine's *Confessions* is conveniently available on line. See http://ccel.org/a/augustine/confessions/.
2. Albert Einstein, *The Expandable Quotable Einstein,* ed. Alice Calaprice (Princeton: Princeton University Press, 2000).
3. Richard P. Feynman, *The Feynman Lectures on Physics* (Reading, MA: Addison-Wesley, 1966), vol. 1, chap. 17: 4.
4. The special theory of relativity was Einstein's first relativity theory. It deals with how things change when we compare the observations of two people moving at a constant velocity relative to each other. The general theory of relativity went farther and attempted to look at how things differ when one of the observers is moving in a gravitational field, specifically, how gravity's ability to accelerate things affects observation. The special theory of relativity is regarded as a branch of the general theory of relativity, in that it applies when the spacetime in which things are moving is considered to be flat—in effect, when no gravity is present. The general theory of relativity shows that gravity affects space and time. See chapter 5.

5. This discrepancy between clocks is known as the *twin paradox* in relativity. It is really not a paradox once you grasp the principle explained here concerning curved and straight trajectories in spacetime.

6. Indeed, in 1894 Albert Michelson, one of the discoverers of the constancy of lightspeed, said that the future of science would only consist of "adding a few decimal places to results already obtained." Michelson believed he was quoting the well-known Lord Kelvin and later confessed he regretted ever having said such a thing.

7. A lightminute is the distance light travels in a minute's time, about 11 million miles. The sun is over 93 million miles from earth, so it takes a little over eight minutes for the sun's rays to reach us.

8. In case you'd like to do the math, the traveler would need to move at a speed relative to us of about 0.999 999 999 999 5 the speed of light. While we wouldn't expect this observer to be earthborn, since it is practically impossible to accelerate any large body to that speed, she could certainly be an extraterrestrial space traveler coming from a distant galaxy that, because of the expansion of the universe, happens to be moving at nearly lightspeed relative to us.

9. In 1896—after flunking the entrance exam on his first try—Einstein became a graduate student at the Eidgenössische Technische Hochschule, in Zurich, Switzerland—called the ETH by all who attended, but known in English as the Swiss Federal Institute of Technology (FIT). It is comparable to the California Institute of Technology (Caltech) in Pasadena and the Massachusetts Institute of Technology (MIT) in Boston.

10. Ronald W. Clark, *Einstein: The Life and Times* (New York: Avon Books, 1972), 160.

CHAPTER 4

1. B. K. S. Iyengar, *Light on the Yoga Sutras of Patanjali* (San Francisco: Thorsons, 1996), 33.

2. See http://www.cosmicharmony.com/Sp/Ramana/Ramana.htm.

3. Self-realization means more than just body awareness. Put simply, it implies a sense of the deeper soul within—a connection to spirit.

4. There are many sources on this remarkable man, who truly realized he was beyond space and time. See, for example, http://www.cosmicharmony.com/Sp/Ramana/Ramana.htm on line or David Goodman. Ed., *Be As You Are: The Teachings of Sri Ramana Maharshi,* (Boston: Arkana, 1985).
5. See "In the Self There is No Space-time" at http://www.cosmicharmony.com/Sp/Ramana/Ramana.htm.
6. Daniel Dennett and Marcel Kinsbourne, "Time and the Observer," *Behavioral and Brain Sciences* 15, no. 2 (1992): 183–247.
7. P. Kolers and M. von Grünau, "Shape and Color in Apparent Motion," *Vision Research* 16 (1976): 329–35.
8. Benjamin Libet, W. Elwood Wright, Bertram Feinstein, and Dennis K. Pearl, "Subjective Referral of the Timing for a Conscious Sensory Experience: A Functional Role for the Somatosensory Specific Projection System in Man," *Brain* 102, no. 1 (1979): 193–224.
9. See Daniel Dennett and Marcel Kinsbourne, "Time and the Observer," *Behavioral and Brain Sciences* 15, no. 2 (1992): 183–247. In this article, the authors attempt to explain the paradoxes using a complex theory involving the mind making multiple drafts of possible scenarios. Just how the mind does this, is not clear. I, too, have attempted to explain Libet's data, using an argument based on quantum physics called the two-time transactional interpretation model. See Fred Alan Wolf, "The Timing of Conscious Experience," *Journal of Scientific Exploration* 12, no. 4 (Winter 1998): 511–42.
10. Mircea Eliade, *Australian Religions: An Introduction* (Ithaca, NY and London: Cornell University Press, 1973), 45.
11. Ibid., 66.

CHAPTER 5

1. Wells wrote this book in 1895, well before airplanes were invented.
2. Tom Van Flandern, a physicist at the University of Maryland, explains that the general theory of relativity predicts that clocks in a stronger gravitational field will tick slower than those in a

weaker field, while the special theory of relativity predicts that moving clocks will tick slower than nonmoving ones. These two effects remarkably manage to cancel themselves out for all earthbound clocks! (See Franco Selleri, ed., *Open Questions in Relativistic Physics* [Montreal: Apeiron, 1998], 81–90; also see http://www.metaresearch.org/cosmology/gps-relativity.asp). Let me explain:

Clocks on the earth are both in a gravitational field and—even though they remain at rest on earth—moving at various rotational speeds, depending on their location on our planet. As the earth daily rotates, every object on it also rotates, but not all at the same speed. The closer the clock is to the axis of rotation, the smaller is its speed. The strength of the field's effect on any global clock depends on how far that clock is from the earth's center. The closer it is to the center, the stronger the gravitational effect. Because the earth's rotation tends to make it bulge at the equator (the earth really isn't a sphere—it's more like a rounded donut without the hole), clocks at the earth's poles tick slower than clocks at the equator. On the other hand, since they lay on the axis of rotation, the polar clocks aren't moving relative to their equatorial counterparts, which rotate at nearly 1,000 mph. Hence the equatorial clocks are slowed by their higher rotational speed relative to their polar counterparts. So even though equatorial clocks should tick faster than polar clocks due to the general theory of relativity, and slower than polar clocks due to the special theory of relativity, the two effects actually cancel each other out. In fact, they cancel each other at any point on the globe, for as the bulging decreases with location north or south of the equator, so does the relative rotational speed.

However, while the canceling effect due to motion and gravity does occur at the earth's surface, the two *don't* cancel each other when you move off the earth or increase your speed. Global positioning satellites carry atomic clocks aboard and use them to locate earthbound objects accurately, but they need adjustment from time to time. The general theory of relativity predicts that these atomic clocks orbiting at an altitude of 16,528 miles will tick faster by about 45,900 nanoseconds/day because they are in a weaker gravitational field than atomic clocks on the earth's surface. The special theory of relativity predicts that these atomic clocks moving at their orbital speeds will tick slower by about 7,200 nanoseconds/day than stationary ground clocks. The net result is a speeding up of the orbiting clocks as compared with earthbound clocks. The highest precision data shows that the on-board atomic clock rates do indeed conform to this prediction and have to be adjusted accordingly.

3. Although the general theory of relativity is quite difficult to work with, its basic equation remains quite simple. It says $T = 8\pi G$, where T stands for the stress produced by energy and mass while G stands for the metric of spacetime—a measure figure that depends on such things as how round or flat space and time appear to be. The factor 8π is just eight times the area of a circle with a radius of one. Just as Einstein's famous $E = mc^2$ equation relating energy to mass says that energy and mass are essentially the same thing, this equation says that energy-mass equals stretch marks on space and time, and therefore is the same thing as spacetime rippling.

4. R. V. Pound and G. A. Rebka, "Apparent Weight of Photons," *Physical Review Letters* 4 (1960): 337–41. The building in which this experiment was carried out was also certainly participating in rotation with the earth. Thereby, all clocks in it were rotating at slightly different speeds, depending on if they were on the top floor or the bottom floor. Nevertheless, the disparity in time made by the gravitational difference was indeed measurable and, although quite tiny, considerably greater than any time shift due to the slight difference in rotational speed.

5. Karl Schwarzschild, "Über das Gravitationsfeld eines Massenpunktes nach der Einsteinschen Theorie," *Sitzbar. Deut. Akad. Wiss. Berlin. Kl. Math.- Phys. Tech.* (1916): 189–96.

6. The black hole's gravity field would pull unequally on the traveler. Assuming the traveler was entering the black hole feet first, the force at her feet would be significantly greater than the force at her head, and this inequality of forces would pull her apart. This difference between the force at her feet and head is similar to the tidal effect produced on our oceans by the moon's gravitational pull and the earth's rotational force.

7. Remember that a lightyear is the vast distance that light travels in one year. Since it goes at slightly greater than 670 million miles per hour, this would mean a distance of just under 6,000 billion miles. Thus a wormhole can make deep space travel as effortless as a walk from one room to another.

8. Kip S. Thorne, *Black Holes and Time Warps: Einstein's Outrageous Legacy* (New York: W. W. Norton & Co., 1994).

CHAPTER 6

1. Michael Dummett, "Causal Loops," in *The Nature of Time,* R. Flood and M. Lockwood, eds. (London: Basil Blackwell, 1986). See also David Deutsch and Michael Lockwood, "The Quantum Physics of Time Travel," *Scientific American* 270, no. 3 (March 1994): 68–74; and Deutsch, "Quantum Mechanics near Closed Timelike Lines," *Physical Review D* 44, no. 10 (November 15, 1991): 3197–3217.

2. You can think of the word *parallel* in a quasi-literal fashion. If you imagine each universe as an infinitely large, flat plane, the two universes reside one atop the other, like pages in a closed book. See Fred Alan Wolf, *Parallel Universes: The Search for Other Worlds* (New York: Simon & Schuster, 1989).

3. Time dilation refers to the slowing down of time for moving clocks as compared with nonmoving clocks. For more on time dilation, see the introduction (remember the muons?) and chapter 3.

4. That time travel does not necessarily violate the chronology tenet was first and convincingly pointed out by physicist David Deutsch in his article, "Quantum Mechanics near Closed Timelike Lines." I tell you more about this in chapter 7.

5. See Kurt Gödel, "An example of a new type of cosmological solution of Einstein's field equations of gravitation," *Reviews of Modern Physics* 21 (1949): 447.

6. The parallel-universes idea has recently been picked up by a new theory of physics called *string theory*, which posits that the universe is made of extremely tiny strings that exist in eleven dimensions. Strings are of two types—open ended and closed. Open-ended strings make up all of the particles of the universe, but closed strings are able to escape one universe and connect up with a parallel universe that string theorists call "branes." Branes are not the same as the parallel universes envisioned in quantum physics prior to string theory, but string theorists like to point out that branes are parallel universes in a higher dimension. I must admit that the idea of closed strings connecting parallel universes does resonate with the ideas present here, but they should not be confused.

7. See David Deutsch, *The Fabric of Reality: The Science of Parallel Universes—and Its Implications* (New York: The Penguin Press, Allen Lane, 1997), 278.

8. Fred Hoyle, *October the First Is Too Late* (New York: Harper & Row, 1966).
9. Everett's ideas explained in simple language can be found in Bryce S. Dewitt, "Quantum Mechanics and Reality," *Physics Today* 23, no. 9 (September 1970): 30–35.
10. Of course, in classical physics when a particular experiment is difficult to control, as for example the trajectory of an air molecule in a room, one would use probability to predict an outcome, such as "the molecule is to the left of a given barrier with a probability of 50 percent."

CHAPTER 7

1. Deutsch, *The Fabric of Reality.* Professor Deutsch founded the idea of using quantum physics to operate computers and is well-known for his efforts. His first paper was published in 1985 and was entitled, "Quantum Theory, the Church-Turing Principle and the Universal Quantum Computer." See *Proceedings of the Royal Society of London*, A 400 (1985): 97–117.
2. Atoms are quite tiny, and usually we, in our everyday lives, need not concern ourselves with them. An atom is so small that if I inflated my thumb, like a balloon, to the size of the earth, one tiny atom of hydrogen contained in one tiny molecule of water that makes up a tiny droplet of perspiration on that thumb would then appear as big as my real thumb.
3. I use the word *possibility* to mean something new. It refers to a quantum physical mathematical quantity that can be imagined to look like a wave issuing from a dropped stone in a calm pond. What makes these waves unusual comes about when the waves reach the pond's boundaries. I call them *possibility*-waves, and I italicize the word *possibility* to remind you that they are a bit mysterious. *Possibility*-waves can do something unimaginable heretofore. They can move through time in either direction. Here we imagine the *possibility*-wave starting from the dropped pebble and, after reaching the shore, suddenly turning around traveling backward through time until it reaches the pebble that started it off. Mathematically speaking, when we multiply the reversed *possibility*-wave with itself, it gives the numerical probability that this particular sequence

of events has. In order for this to make sense, these *possibilities* must be complex numbers. That means they can be positive, negative, and also imaginary (multiplied by the square root of minus one). The two *possibility*-waves then produce a real positive number that describes the actual probability of the two events being connected. While this may seem bizarre, today's radio and TV transmission technologies are based on a very similar way of thinking. When the TV wave, for example, arrives at your TV set, the wave creates a mirror wave (not backward in time, though) in the receptor that multiplies the original wave. From that multiplication the information in the TV wave can be accessed. When quantum waves multiply in this manner, time itself appears.

4 See Nick Herbert, *Quantum Reality* (New York: Anchor Press/ Doubleday), 1985.

5. The world of quantum-physics possibilities is called a *Hilbert space* and represents all of the possibilities as if they were arrows pointing in as many directions as there are dimensions of the space. A two-sided coin would have a two-dimensional Hilbert space, while a particle passing through a screen with, say, 20 slits would have a 20-dimensional Hilbert space.

6. Deutsch, *The Fabric of Reality*.

7. This paper published in 1991 came after Deutsch's seminal 1985 paper on quantum computers. See David Deutsch, "Quantum Mechanics near Closed Timelike Lines," *Physical Review D* 44, no. 10 (November 15, 1991): 3197–3217.

8. For a popular accounting of how quantum computers work, see Julian Brown, *Minds, Machines, and the Multiverse: The Quest for the Quantum Computer* (New York: Simon & Schuster, 2000).

9. Yakir Aharonov, Jeeva Anandan, Sandu Popescu, and Lev Vaidman, "Superpositions of Time Evolutions of a Quantum System and a Quantum Time-Translation Machine," *Physical Review Letters* 64, no. 25 (June 18, 1990): 2965–68.

10. Lev Vaidman, "A Quantum Time Machine," *Foundations of Physics* 21, no. 8 (1991): 947–58.

11. You may be wondering about what the difference is between a gravitational field and a gravitational potential field. A gravity field exerts a force on any object possessing a mass. A gravitational potential field does not. For a force of gravity to arise, a difference in gravitational potential must exist. You can think of an electrical

circuit to see how this difference works. The electrical potential difference present between the plates of the battery is what gives rise to that field of force. When you put a battery into a flashlight and turn the light on, electrical charges move due to the electrical field they feel. As the potential difference wanes, the battery weakens; when it finally disappears, no current flows at all. In the sphere, our time traveler feels no gravity force at all—even though he sits in a strong gravitational potential—because no difference in potential exists anywhere inside the sphere.

CHAPTER 8

1. Although, as we see later in the chapter, negative *possibility*-waves make sense when we square them, negative probabilities make no sense in our usual understanding of the world and have no bearing here. Nevertheless, the concept does show up in the strange realm of quantum physics. So long as the final outcome of any calculation has a positive probability, it is even possible for probabilities to have negative values before the outcome is realized. See Richard P. Feynman, "Negative Probability," in *Quantum Implications: Essays in Honour of David Bohm,*" B. J. Hiley and F. David Peat, eds. (London and New York: Routledge and Kegan Paul, 1987), 235–48.

2. Eventually it was realized that to really represent *possibility*-waves, *complex numbers* had to be used. These are numbers that have two parts: one part is a real positive or negative number, a; and the other part is an *imaginary* number, also either positive or negative, ib. A *possibility*-wave is best represented by the complex number $(a+ib)$. The letter i represents an imaginary number whose square equals minus one. Hence $i = \sqrt{-1}$. As you can surmise, imaginary numbers do not exist in the "real" world. When any imaginary number is multiplied by itself (squared) it always results in a negative number, regardless of whether b is positive or negative. When real numbers are squared, whether they are positive or negative, they always produce a positive number.

3. Actually, it is not true that the square of the *possibility*-wave always gave a positive number, though it is usually stated this way in popular books on quantum physics. It yields a positive number if the *possibility*-wave is represented by a real number, a, and the

probability-curve is represented by the positive number a^2. But if the *possibility*-wave is represented by a complex number $(a+ib)$, then its correct "square" would actually be $(a+ib) \times (a-ib) = a^2 + b^2$, which is always a positive number. (You can do the math if you'd like). The number $(a-ib)$ is called the *complex conjugate* number of $(a+ib)$ and, if you think about it, this "square" makes perfect sense. Simply squaring a complex number as $(a+ib) \times (a+ib) = a^2 - b^2 + 2iab$ yields just another complex number that can't ever be real and positive, unless $b = 0$. You can think of the complex conjugate *possibility*-wave as a mirror image of the original *possibility*-wave.

4. John G. Cramer, "Generalized Absorber Theory and the Einstein-Podolsky-Rosen Paradox," *Physical Review D* 22 (1980): 362. Also see Cramer, "The Transactional Interpretation of Quantum Mechanics," *Reviews of Modern Physics* 58, no. 3 (July 1986).

5. See notes 3 and 4 for this chapter. A complex-conjugate wave is a time-reversed mirror image of the original wave. This turns out to be more than just poetic metaphor, since the waves have similar forms but travel in opposite directions in time, just as a mirror image produces lightwaves that travel in the opposite direction relative to the original waves they reflect.

6. The situation with the complex-conjugate wave is not the first in which someone has noticed that running the clock backwards in a physics equation could lead to a new discovery. Richard Feynman received the Nobel Prize for his use of this idea in the study of the interactions of photons and electrons called *quantum electrodynamics.*

7. I call the complex-conjugate *possibility*-wave a *star wave* in my previous book by the same title. See Fred A Wolf, *Star Wave: Mind, Consciousness, and Quantum Physics* (New York: Macmillan, 1984).

CHAPTER 9

1. Indeed, probability-curves are integral in testing a product, particularly a new medicine, before the product is released to the public. The medicine is tested on a number of individuals, and the drug company expects the test to produce a bell-shaped curve telling them that their product is effective.

2. See chapter 5 of my earlier book, *Matter into Feeling: A New Alchemy of Science and Spirit* (Portsmouth, NH: Moment Point Press, 2002).

3. A similar representation, illustrating the spread of a *possibility*-wave and its focusing upon measurement of a subatomic particle, was first used in an excellent book by physicist Robert H. March. See March, *Physics for Poets* (New York: McGraw-Hill Book Co., 1970), 228.

4. Amit Goswami, *Quantum Mechanics* (Dubuque, IA: Wm. C. Brown, 1992), 517–23. Also see Goswami, *The Self-Aware Universe: How Consciousness Creates the Material World* (New York: Tarcher/Putnam, 1993).

5. See Roger Penrose, *Shadows of the Mind: A Search for the Missing Science of Consciousness* (New York: Oxford University Press, 1994). Also see Penrose, *The Emperor's New Mind: Concerning Computers, Minds, and the Laws of Physics* (New York: Oxford University Press, 1989).

6. See John G. Cramer, "The Transactional Interpretation of Quantum Mechanics," *Reviews of Modern Physics* 58, no. 3 (July 1986).

CHAPTER 10

1. Taken from "Viewpoints on String Theory" by Sheldon Glashow and from transcripts of the television show "The Elegant Universe" shown on PBS during the fall, 2003. See the web page: http://www.pbs.org/wgbh/nova/elegant/view-glashow.html for more. Written, produced, and directed by Julia Cort and Joseph McMaster; Series Producer and Director, David Hickman.

2. A current theme in science today proposes that consciousness arises from material processes, not the other way around. Hence, the main proponents of this belief see no role for consciousness in the evolution of the universe. They would say that since human beings weren't around in the distant past, human consciousness could not influence the big bang or the current cosmology, based as it is on the past. I counter this concept with the thought that consciousness need not be human. Hence, I refer to the idea of a Mind of God, which has recently been used in several popular books.

3. For examples of how time travel appears in psychology, see Daniel C. Dennett and Marcel Kinsbourne, "Time and the Observer," *Behavioral and Brain Sciences* 15, no. 2 (1992): 183–247. In physics, see David Deutsch, "Quantum Mechanics near Closed Timelike Lines," *Physical Review D* 44, no. 10 (November 15, 1991): 3197–3217. In ancient spiritual practice, see B. K. S Iyengar, *Light on the Yoga Sutras of Patanjali* (San Francisco: Thorsons, 1996), 37.

4. See Fred Alan Wolf, *Parallel Universes: The Search for Other Worlds* (New York: Simon & Schuster, 1989), 225–233. Also see Avshalom C. Elitzur, Shahar Dolev, and Anton Zeilinger, "Time-Reversed EPR and the Choice of Histories in Quantum Mechanics," available on line at http://arxiv.org/abs/quant-ph?0205182.

5. See Fred Hoyle, "The Universe: Past and Present Reflections" (Preprint Series no. 70, Department of Applied Mathematics and Astronomy, University College, Cardiff, U.K., May 1981). In this article Hoyle discusses all the different memories created by parallel selves following alternate paths. He points out that ". . . we would then have no means within the [parallel-universes] theory of specifying the particular route of which we are consciously aware. To treat all routes equally, one would need to postulate an ensemble of alter egos who are consciously aware of the other routes. Since each route satisfies the dynamical equations (including the interactions) quite independently of the other routes, there would be no way to compare notes with an alter ego, so at least in this respect the situation would be free from contradiction."

   Further in the article he says, "In the ensemble of our lives, with repeated switches to the memory sequences of our alter egos, there would be all the existences we would have experienced if our snap decisions had been differently taken. There could be snap decisions affecting our behavior in moments of danger, the places we visit, the people we meet, and perhaps even the people we marry. There is the possibility of waking each morning beside a different spouse, although our memory each morning will always be consistent with the spouse-of-the-day, and we will therefore be entirely unaware of the other possibilities."

6. For reduction in crime statistics in Washington D.C. by using Transcendental Meditation, see http://mum.edu/tm_research/tm_dc/Abstract.html. For reversal of aging process by usingTranscendental Meditation, see http://mum.edu/tm_research/tm_charts/5MVAH_A.html.

Chapter 11

1. Eknath Easwaran, trans. *The Bhagavad Gita* (Tomalas, CA: Nilgiri Press, 1985), 37.
2. See Amit Goswami, *The Self-Aware Universe: How Consciousness Creates the Material World* (New York: Tarcher/Putnam, 1993); Roger Penrose, *Shadows of the Mind: A Search for the Missing Science of Consciousness* (New York: Oxford University Press, 1994); and H. P. Stapp, *Mind, Matter, and Quantum Mechanics* (New York: Springer-Verlag, 1993/2003).
3. This quote comes from a four-part program, *Death: The Trip of a Lifetime,* shown on PBS in October 1993. It was produced, written, and hosted by Greg Palmer and produced by Sue McLaughlin as part of a KCTS-Seattle/TV-New Zealand/Australian Broadcasting Commission production.
4. See http://reluctant-messenger.com/yoga-sutras-4.htm and B. K. S Iyengar, *Light on the Yoga Sutras of Patanjali* (San Francisco: Thorsons, HarperCollins, 1996).

BIBLIOGRAPHY

Adejumo, Ebenezer Ademola. "The Concept of Time in Yoruba, Australian Aboriginal, and Western Cultures, Especially as It Is Manifested in the Visual Arts." M.S. thesis, Flinders University of South Australia, 1976.

Aharonov, Yakir, Jeeva Anandan, Sandu Popescu, and Lev Vaidman. "Superpositions of Time Evolutions of a Quantum System and a Quantum Time-Translation Machine." *Physical Review Letters* 64, no. 25 (June 18, 1990): 2965–68.

Aharonov, Yakir, Peter G. Bergmann, and Joel L. Lebowitz. "Time Symmetry in the Quantum Process of Measurement," *Physical Review* 134B (1964): 1410–16.

Albert, David Z. "A Quantum Mechanical Automaton." *Philosophy of Science* 54 (1987): 577–85.

———. *Time and Chance.* Cambridge, MA: Harvard University Press, 2000.

Békésy, George von. *Sensory Inhibition.* Princeton, NJ: Princeton University Press, 1967.

Brown, Julian. *Minds, Machines, and the Multiverse: The Quest for the Quantum Computer.* New York: Simon & Schuster, 2000.

Clark, Ronald W. *Einstein: The Life and Times.* New York: Avon Books, 1972.

Colin, Dean. "The Australian Aboriginal "Dreamtime": An Account of Its History, Cosmogenesis, Cosmology," and Ontology." B.S. thesis, Deakin University, 1990. Available from the Australian Institute of Aboriginal Studies, Canberra.

Cramer, John G. "Generalized Absorber Theory and the Einstein-Podolsky-Rosen Paradox." *Physical Review D* 22 (1980): 362.

———. "The Transactional Interpretation of Quantum Mechanics." *Reviews of Modern Physics* 58, no. 3 (July 1986): 647–87.

Davies, Paul. *About Time.* New York: Touchstone, 1996.

Dennett, Daniel C., and Marcel Kinsbourne. "Time and the Observer." *Behavioral and Brain Sciences* 15, no. 2 (1992): 183–247.

Deutsch, David. *The Fabric of Reality: The Science of Parallel Universes—and Its Implications.* New York: Penguin Press, Allen Lane, 1997.

———. "Quantum Mechanics near Closed Timelike Lines." *Physical Review D* 44, no. 10 (November 15,1991): 3197–3217.

———. "Quantum Theory, the Church-Turing Principle and the Universal Quantum Computer." Proceedings of the Royal Society of London, Series A 400 (1985): 97–117.

Dewitt, Bryce S. "Quantum Mechanics and Reality," *Physics Today* 23, no. 9 (September 1970): 30–35.

Easwaren, Eknath, trans. *The Bhagavad Gita.* Tomalas, CA: Nilgiri Press, 1985.

Einstein, Albert. *The Expandable Quotable Einstein.* Edited by Alice Calaprice. Princeton, NJ: Princeton University Press, 2000.

Eliade, Mircea. *Australian Religions: An Introduction.* Ithaca and London: Cornell University Press, 1973.

Elitzur, Avshalom C., Shahar Dolev, and Anton Zeilinger. "Time-Reversed EPR and the Choice of Histories in Quantum Mechanics." Paper presented at the Solvay Conference in Physics, Delphi, Greece, November 2001. Available online at http://arxiv.org/abs/quant-ph?0205182.

Feuerstein, Georg. *Tantra: The Path of Ecstasy.* Boston and London: Shambala, 1988.

Feynman, Richard P. *The Feynman Lectures on Physics.* Vol. 1. Reading, MA: Addison-Wesley, 1966.

Feynman, Richard P. "Negative Probability." In *Quantum Implications: Essays in Honour of David Bohm.* Edited by B. J. Hiley and F. David Peat. London and New York: Routledge and Kegan Paul, 1987.

Goswami, Amit. *Quantum Mechanics.* Dubuque, IA: Wm. C. Brown, 1992.

———. *The Self-Aware Universe: How Consciousness Creates the Material World.* New York: Tarcher Putnam, 1993.

Herbert, Nick. *Quantum Reality.* New York: Anchor Press/Doubleday. 1985.

Hoyle, Fred. *The Intelligent Universe.* New York: Holt, Rinehart, & Winston, 1983.

———. *October the First is Too Late.* New York: Harper & Row, 1966.

———. "The Universe: Past and Present Reflections." Preprint Series no. 70, Department of Applied Mathematics and Astronomy, University College, Cardiff, U.K., May 1981.

Hoyle, F., and J. V. Narliker. *Action at a Distance in Physics and Cosmology.* San Francisco: W. H. Freeman & Co., 1974.

Iyengar, B. K. S. *Light on the Yoga Sutras of Patanjali.* San Francisco: Thorsons, 1996.

Kezwer, Glen Peter. *Meditation, Oneness, and Physics.* New York: Lantern Books, 2003.

Kolers, P., and M. von Grünau. "Shape and Color in Apparent Motion." *Vision Research* 16 (1976): 329–35.

Libet, B., E. W. Wright, B. Feinstein, and Dennis Pearl. "Subjective Referral of the Timing for a Conscious Sensory Experience: A Functional Role for the Somatosensory Specific Projection System in Man." *Brain* 102, no. 1 (March 1979): 193–204.

Love, W. "Was the Dream-Time Ever a Real-Time?" *Anthropological Society of Queensland Newsletter* 196 (1989): 1–14. Available from the Australian Institute of Aboriginal Studies.

March, Robert H. *Physics for Poets.* New York: McGraw-Hill, 1970.

Penfield, Wilder. *The Mysteries of the Mind.* Princeton, NJ: Princeton University Press, 1975.

Penrose, Roger. *The Emperor's New Mind: Concerning Computers, Minds, and the Laws of Physics.* New York: Oxford University Press, 1989.

———. *Shadows of the Mind: A Search for the Missing Science of Consciousness.* New York: Oxford University Press, 1994.

Pouley, Jim. *The Secret of Dreaming.* Templestowe, Australia: Red Hen Enterprises, 1988.

Pound, R. V., and G. A. Rebka. "Apparent Weight of Photons." *Physical Review Letters* 4 (1960): 337–41.

Prabhupada, A. C. Bhaktivedanta Swami. *Bhagavad Gita as It Is*, 11:7, in *The Library of Vedic Culture*. Interactive CD. Los Angeles: The Bhaktivedanta Book Trust, 2003.

Price, Huw. *Time's Arrow and Archimedes' Point.* New York: Oxford University Press. 1996.

Rossi, B., and D. B. Hall. "Variation of the Rate of Decay of Mesotrons with Momentum." *Physical Review* 59 (1941): 223–28.

Saint Augustine, *Confessions.* Complete text available on line at http://ccel.org/a/augustine/confessions/.

Schwarzschild, Karl. "Über das Gravitationsfeld eines Massenpunktes nach der Einsteinschen Theorie." (On the Gravitational Field of a Mass-point according to Einstein's Theory.) *Sitzbar. Deut. Akad. Wiss. Berlin. Kl. Math.-Phys. Tech.* [Preface §§14.1, 23.6] (1916): 189–96.

Shanley, William, ed. *Alice zwischen den Welten.* (Alice between the worlds). Stuttgart: Deutsche Verlags-Anstalt, 1999.

Spencer, B., and F. J. Gillen. *The Native Tribes of Central Australia.* New York: Dover, 1968 (first published in 1899).

Stanner, W. H. *White Man Got No Dreaming: Essays 1938–1973.* Canberra, Australia: Australian National University Press, 1979.

Stapp, H. P. *Mind, Matter, and Quantum Mechanics.* New York: Springer-Verlag, 1993/2003.

———. "Quantum Theory and the Role of Mind in Nature." *Foundations of Physics* 11 (2001): 1465–99.

Sutton, Peter, ed. *Dreamings: The Art of Aboriginal Australia.* Victoria, Australia: Penguin Books, 1988.

Thorne, Kip S. *Black Holes and Time Warps: Einstein's Outrageous Legacy*. New York: W. W. Norton & Co., 1994.

Tipler, Frank J. "Rotating Cylinders and the Possibility of Global Causality Violation." *Physical Review D* 9 (1974): 2203.

Vaidman, Lev. "A Quantum Time Machine." *Foundations of Physics* 21, no. 8 (1991): 947–58.

Van Flandern, Tom. *Open Questions in Relativistic Physics.* Edited by Franco Selleri. Montreal: Apeiron, 1998.

Visser, M., S. Kar, and N. Dadhich. "Traversable Wormholes with Arbitrarily Small Energy Condition Violations." *Physical Review Letters* vol. 90, electronic reference 201102 (2003).

Wolf, Fred Alan. *The Dreaming Universe: A Mind-Expanding Journey into the Realm Where Psyche and Physics Meet.* New York: Simon and Schuster, 1994. Reprint, New York: Touchstone, 1995.

———. *Matter into Feeling: A New Alchemy of Science and Spirit.* Portsmouth, NH: Moment Point Press, 2002.

———. "On the Quantum Physical Theory of Subjective Antedating." *Journal of Theoretical Biology* 136 (1989): 13–19.

———. *Parallel Universes: The Search for Other Worlds.* New York: Simon & Schuster, 1989.

———. *Star Wave: Mind, Consciousness, and Quantum Physics.* New York: Macmillan, 1984.

———. *Taking the Quantum Leap: The New Physics for Nonscientists.* San Francisco: Harper & Row, 1981. Revised edition, New York: HarperCollins, 1989.

———. "The Timing of Conscious Experience." *Journal of Scientific Exploration* 12, no. 4 (Winter 1998): 511–42.

Yogananda, Paramahansa. *Autobiography of a Yogi.* Los Angeles: Self Realization Fellowship, 1973.

INDEX

QUEST BOOKS
encourages deep and open-minded inquiry into
world religions, philosophy, science, and the arts
in order to understand the wisdom of the ages,
realize the unity of all life, and help people explore
individual spiritual self-transformation.

Its publications are generously supported by
The Kern Foundation,
a trust committed to Theosophical education.

Quest Books is the imprint of
The Theosophical Publishing House,
a division of The Theosophical Society in America.

For information about programs, literature,
on-line study, international centers, and
membership benefits, see
www.theosophical.org
or call 800-669-1571 or
(outside the U.S.) 630-668-1571.

Related Quest titles:

The Visionary Window, Amit Goswami
Science and the Sacred, Ravi Ravindra
Beyond the Postmodern Mind, Huston Smith
The Wisdom of the Vedas, J. C. Chatterji
Two Faces of Time, Lawrence Fagg
The Play of Consciousness in the Web of the Universe, Edward L. Gardner

ABOUT THE AUTHOR

Fred Alan Wolf, Ph.D., is the author of eleven books and universally acclaimed for his simplification of the new physics in a way that integrates its findings with world spiritual traditions and contemporary consciousness studies. One of his best-known titles is *Taking the Quantum Leap,* which won the prestigious National Book Award for Science in 1982.

Dr. Wolf earned his doctorate in theoretical physics from the University of California at Los Angeles in 1963. Since then, his investigations have taken him from intimate discussions with physicist David Bohm to the mysterious jungles of Peru; from master classes with Nobel Laureate Richard Feynman to shamanic journeys on the high deserts of Mexico; from a significant meeting with physicist Werner Heisenberg to the hot coals of a fire walk.

A former professor of physics at San Diego State University, Dr. Wolf has also taught at the University of London, the University of Paris, the Hahn-Meitner Institute for Nuclear Physics in Berlin, the Hebrew University of Jerusalem, the Philosophical Research Society, and the Holmes Institute. He is an honored member of the Martin Luther King, Jr. Collegium of Scholars.

Dr. Wolf is currently featured in a major motion picture, *What the Bleep Do We Know?!,* and has also appeared as the resident physicist on the Discovery Channel's "The Know Zone" and on many radio and television programs in the United States and abroad. He lectures and teaches worldwide and lives in San Francisco.

To contact Dr. Wolf, write c/o Quest Books, P. O. Box 270, Wheaton, IL 60189-0270, or email him at fred@fredalanwolf.com. Visit his web page at **www.fredalanwolf.com**.